Chironomidae of Central America

Chironomidae of Central America

An Illustrated Introduction to Larval Subfossils

Ladislav Hamerlík and Fabio Laurindo da Silva

CRC Press is an imprint of the
Taylor & Francis Group, an **informa** business

First edition published 2021
by CRC Press
6000 Broken Sound Parkway NW, Suite 300, Boca Raton, FL 33487-2742

and by CRC Press
2 Park Square, Milton Park, Abingdon, Oxon, OX14 4RN

CRC Press is an imprint of Taylor & Francis Group, LLC

Library of Congress Control Number: 2020942347

ISBN: 978-0-367-55821-5 (hbk)
ISBN: 978-0-367-07606-1 (pbk)
ISBN: 978-0-429-02157-2 (ebk)

DOI: 10.1201/9780429021572

Typeset in Times
by Deanta Global Publishing Services, Chennai, India

Contents

Preface

Larvae of tiny insects belonging to the family Chironomidae, known as non-biting midges, chironomids, and bloodworms, are powerful indicators of environmental conditions. Not only do they live all over our planet, colonizing virtually all types of aquatic ecosystems from the smallest ponds to the deepest lakes, and have thousands of species with various ecological requirements, but after they die, their head capsules remain in lake sediments for thousands of years. After obtaining a sample of the sediment, their remains can be used to reconstruct the history of the lake and its surroundings. For reliable paleolimnological reconstructions, accurate identification of subfossil remains is essential, since with misidentified subfossil material, even the most comprehensive study will be meaningless. However, for a long period, reliable identification of the remains was rather problematic because the existing identification literature focused on identification of living insects, which possess significantly more identification features relative to dramatically worn down subfossil remains with a minimum of such structures.

The situation changed markedly with the publication of the paleokey by Brooks et al. (2007). This comprehensive guide summarized and illustrated 198 morphotypes from 112 genera of 6 subfamilies taken from lake sediments mainly deposited all over Europe. Moreover, they reviewed the knowledge on chironomid taxonomy, ecology, and paleolimnology, as well as field and laboratory techniques. It is not an exaggeration to say that this guide is essential for all "dead head" specialists; there are plenty of (using Ian Walker's words) "well-abused, coffee-stained and dog-eared" copies on laboratory desks all over the world (including the authors'). Since most of the featured taxa in the guide are widespread, it is relevant also in other parts of the world. However, when it comes to other biogeographical regions with a large number of endemic taxa, such as the Neotropical region, a Palaearctic guide may have limited applicability.

Paleolimnological investigations using chironomid remains in Central America and adjacent areas have not been extensive, but lately have undergone a resurgence of interest with many workers following the success of chironomid-based reconstructions in many parts of the world.

This highlights the need for a taxonomic guide that will ultimately be helpful for a thorough analysis of the diversity and distribution of the taxa encountered to date in the region. In this context, we decided to publish this identification guide, illustrating the main chironomid morphotypes found in lake-surface sediments and sediment cores in Northern Central America. Of the total 64 genera found, it includes 20 endemic ones not listed in Brooks et al. (2007) and many new/different morphotypes from the listed genera. We hope that this guide will be useful for people working with subfossil material, not only in Central America, but in the whole Neotropical region.

Finally, this book is a result of a joint effort of a (paleo)limnologist and a taxonomist. We are convinced that a reliable subfossil guide cannot be developed without a thorough taxonomical knowledge of the region's recent chironomid fauna. At the same time, using a paleolimnological approach, it is possible to transmit this information to morphotypes that can be linked with ecology and, reaching beyond taxonomy and limnology, used to reconstruct the past development of nature. After all, paleo-workers and taxonomists have the same goal: to learn more about these fascinating insects and, through them, discover the world around us.

Authors

Ladislav Hamerlík, PhD, is an aquatic ecologist, fascinated by the variability and diversity of Chironomidae. He studies ecology and taxonomy of these tiny insects, mainly, as indicators of both recent and past environmental changes all over the world. Currently, he holds the position of associate professor in the Department of Biology and Ecology, Faculty of Natural Sciences, Matej Bel University, Banská Bystrica, and at the Institute of Zoology, Slovak Academy of Sciences, Bratislava, Slovakia.

Fabio Laurindo da Silva, PhD, is currently professor in the Department of Zoology, Institute of Biosciences, University of São Paulo, São Paulo, Brazil, where he teaches courses on biogeography, aquatic insects, and general entomology and conducts research on evolutionary biology, historical biogeography, and systematics of Chironomidae.

Acknowledgments

This book would not have been possible without the help and contribution of the following people: Antje Schwalb, Sergio Cohuo Duran, Laura Macario-Gonzales, Liseth Pérez, Julieta Massaferro, Marta Wojewódka, Edyta Zawisza, and Krystyna Szeroczyńska, who provided us with materials from Central American lakes. We are grateful to Martin Spies, Torbjørn Ekrem, John Epler, Julieta Massaferro, Luiz Carlos Pinho, Susana Trivinho-Strixino, Sofia Wiedenbrug, and the Chironomid Exchange Forum for help with the identification of some "mysterious midges." We are also grateful to John Epler for his thoughtful review of the early draft of this book. Some of the figures presented in the book were kindly provided by P.S. Cranston. Peter Bitušík's comments on the manuscript are also greatly acknowledged.

Finally, LH is grateful to Alina for her endless patience and support and Hanna for the motivation to finish the book. LH was funded by the National Science Centre, Poland, Contract No. 2015/19/P/ST10/04048. The project received funding from the European Union's Horizon 2020 Research and Innovation Program, under the Marie Skłodowska-Curie Grant Agreement No. 665778.

FLS was supported by fellowships from the Coordination for the Improvement of Higher Education Personnel (CAPES: 2014/9239-13-8) and São Paulo Research Foundation (FAPESP: 2016/07039-8 and 2018/01507-5).

CHIRONOMIDAE IN A NUTSHELL

Non-biting midges (Diptera, Chironomidae) are the most widely distributed free-living holometabolous insects, with larvae occurring in most aquatic and semiaquatic habitats, and to a lesser extent in brackish and marine waters, and semi-terrestrial and terrestrial ecosystems (Oliver 1971). The range of environmental conditions under which chironomids are found is more extensive than that of any other family of aquatic insects (Silva et al. 2018). They often dominate aquatic insect communities in both abundance and species richness (e.g., Cranston 1982; Ferrington 2008) and compose a significant fraction of the macrozoobenthos of most freshwater ecosystems. Oliver (1971) and Armitage et al. (1995) have provided a sufficient overview of chironomid biology and ecology. Their great species and habitat diversity, combined with ecological adaptability, makes these insects an important environmental indicator for aquatic ecosystems.

The family Chironomidae possibly originated in the middle Triassic, around 248–210 million years ago (Cranston et al. 2010). The group includes 11 subfamilies (Aphroteniinae, Buchonomyiinae, Chironominae, Chilenomyiinae, Diamesinae, Orthocladiinae, Podonominae, Prodiamesinae, Tanypodinae, Telmatogetoninae, and Usambaromyiinae) and comprises at least 10,000 species in more than 400 genera (Armitage et al. 1995; Sæther 2000). This high species diversity has been attributed to the antiquity of the family, with rather low vagility, leading to isolation and evolutionary plasticity (Armitage et al. 1995). Approximately 900 species of Chironomidae are recognized from the Neotropical region (M. Spies, pers. comm.), of which most are assigned to one of the three largest subfamilies: Chironominae, Orthocladiinae, and Tanypodinae. The remaining subfamilies are, in general, less species-rich and have one or only few species within one or few genera (Spies et al. 2009).

Compared with other insects, chironomid larvae are well preserved in lake sediments for a long period due to their chitinized head capsules, which allow their remains to be used for both limnological and paleolimnological studies. As lakes accumulate organic and inorganic remains constantly since their formation, their sediments represent a continuous environmental archive that holds information about the past of the lake and its environment. Moreover, lake-surface sediments, accumulating in the deepest parts of the lake, can also be important sources of recent distribution of species, because they contain remains of the contemporary biota from different parts of the system (Frey 1976). While recent years have seen increased activity regarding Chironomidae in the Neotropical region (e.g., Watson and Heyn 1992; Silva and Gessner 2009; Andersen et al. 2015; Siri 2015; Trivinho-Strixino et al. 2013, 2015; Parise and Pinho 2016; Silva and Oliveira 2016; Epler 2017; Silva and Ferrington 2018; Dantas et al. 2020), the knowledge of paleoenvironmental archives

DOI: 10.1201/9780429021572-1

preserved in lake sediments, particularly in Central America and the Caribbean region, remains fragmentary and limited to only a few studies (Vinogradova and Riss 2007; Pérez et al. 2010, 2013; Hamerlík et al. 2018a,b; Hamerlík and Silva 2018; Wu et al. 2015, 2017, 2019a,b).

GEOGRAPHICAL RANGE

Central America is a region of particular interest for ecological and biogeographical studies, because it represents a bridge between two main biogeographical realms, the Nearctic and the Neotropical regions. In spite of its reduced area, the equatorial position, complex topography, and range of local weather patterns produce numerous microhabitats that support one of the largest global biodiversity hotspots, the Mesoamerican forest, which extends from northern Guatemala to central Panama. Although this biome is one of the most endangered ecosystems in the Neotropics (Sánchez-Azofeifa et al. 2014) due to high rates of habitat loss and fragmentation (Chacon 2005), there is still considerable opportunity for conservation action (Albuquerque et al. 2015). In this context, lake sediments may be a valuable source of information, not only of past environmental changes but also of recent distribution of species, since the sediment accumulating in the deepest part of a lake represents the mixture of the biological community from different parts of the lake and its surroundings, as well as accumulation of communities from different time periods (Hamerlík et al. 2018b).

Several biogeographic investigations indicate that the Neotropical region is composed of several subregions with a divergent evolutionary history, showing relationships with different continents in the past (Cione et al. 2015; Silva and Farrell, 2017). Therefore, the biogeographical coverage here adopts limits to the Neotropical region (sensu Morrone 2014), focusing mainly on Central America. This area does not include the Neotropical Transition Zone and Andean-Patagonian units, and their endemic genera are not covered herein.

PURPOSE AND SCOPE

Identification of larval Chironomidae tends to be complicated due to a lack of diagnostic morphological features to distinguish several genera. Moreover, identification of some species can only be achieved by individual rearing of larvae and collecting larval and pupal skins to establish the associations between life stages (Silva and Wiedenbrug 2014). Therefore, without larval rearing, genus identification of isolated immature stages must often be considered as tentative (Spies et al. 2009). Consequently, a key to living chironomids can be offered here only partly, since many records for Central America are based only on adult male specimens. On the other hand, the key to subfossil chironomids is fully illustrated and includes all the genera that were examined from sediment samples, from lakes across Central America (for review, see Hamerlík et al. 2018b, 2020).

Before a chironomid head capsule becomes buried in the sediment, it is part of a larva of a living individual. The head of a living larva contains a myriad of minute structures, such as organelles, palps, setae, lamellae, etc., that are often essential for distinguishing different taxa. When a larva dies or goes through molting, the hard-sclerotized remains of the body are buried in the sediment, while the rest of the body disintegrates. Moreover, due to taphonomical processes and mechanical damage over time, several fragile structures are lost and the diagnostic characters are usually reduced to the basic, most robust structures, such as the mentum and mandibles. Reduced diagnostic characters lead to limited potential for accurate identification of remains and require specific identification keys focusing only on the basic characters. The applicability of this approach, called parataxonomy, is limited, but in some fields, it can be the only way to get results on highly important scientific questions (Krell 2004). In surface sediments, both living larva and subfossil remains can be encountered. Thus, in this book, we provide two identification keys: (1) to living chironomids, which can be used to identify larvae bearing all the important diagnostic features and (2) to subfossil chironomids, which aims to distinguish larval subfossil remains containing only a limited number of characters.

Paleolimnology is the science that uses the physical, chemical, and biological information preserved in lake sediments to reconstruct past environmental changes in inland aquatic ecosystems. This subject area includes knowledge from a number of diverse fields of study, such as limnology, geology and ecology, but because of challenges that separate researchers from direct knowledge on past lake conditions, the discipline is multidisciplinary by necessity (Whitmore and Riedinger-Whitmore 2014). In light of this, we hope this book will be useful for students, investigators, and other professionals working in Central America and will bridge the gap between taxonomists, limnologists and paleolimnologists.

TERMINOLOGY AND MORPHOLOGY

Chironomidae are nematocerous dipterans with four larval instars. Even though most structures seem to be present in earlier instars, morphologic and taxonomic examinations are made on the final instar (Cranston 1995). This is because several traits of the latter, particularly ratios and shapes, do not apply to earlier instars and do not permit reliable identifications (ibid.). The great ecological variability displayed by Chironomidae species is related to the wide array of morphologic adaptations present among members of the group (Ferrington et al. 2008). Cranston (1995, 2013) and Epler (2001) diagnosed the fourth instar chironomid larva as follows: larvae possess an exposed, complete, sclerotized, non-retractable cephalic capsule and an elongated, cylindrical body, lacking jointed thoracic legs. The antenna is well developed and multi-segmented (in Tanypodinae, capable of complete retraction into head); in some taxa (especially tribe Tanytarsini), it may be mounted on a tubercle. Mouthparts of larval chironomids are directed anteriorly, with opposing mandibles operating ventromedially with either a transverse, usually toothed plate (called a mentum), or a movable, toothed ligula (in Tanypodinae). The body consists of three thoracic and nine abdominal segments; an anterior and a posterior pair of parapods bearing spines or claws are usually present. The last body segment generally bears paired procerci and four anal papillae, in some species, also short, lateral and/or longer, ventral tubuli.

Several different terminologies have been in fairly frequent usage, with some being consistent with prevailing views of insect morphology, while others have been more idiosyncratic (Cranston 1995). Since the publication of Sæther (1980), a detailed evaluation of morphologic homology, most discussions of morphologic structures have used the terms included in that glossary (Spies et al. 2009).

KEY TO THE LARVAE OF THE GENERA OF CHIRONOMIDAE OF THE NEOTROPICAL REGION

Genera that are illustrated in the book as subfossil remains are marked in **bold**.

1 Antenna capable of complete retraction into head. Prementum with a distinctive ligula bearing 4–8 teeth. Mentum weakly developed, partially membranous with comb. TANYPODINAE 10

1' Antenna not retractile. Prementum variously developed, ligula never developed as in Tanypodinae. Mentum usually well developed and sclerotized 2

DOI: 10.1201/9780429021572-2

2(1) Ventral part of mentum laterally developed into ventromental plates, usually striate and never with beard beneath. CHIRONOMINAE..33

2' Ventromental plates, if developed, never striated (some *Nanocladius* may have a few markings on ventromental plates) and sometimes with beard beneath..3

3(2) Mentum untoothed. Anterior labrum with elongated sensorial setae. Body covered with sclerotized plates and strong setae (except for *Paraphrotenia*).........................APHROTENIINAE

3' Mentum nearly always toothed. Anterior labrum with short sensorial setae. Body smooth, although may be strongly setose....4

4(3) Premandible absent...5

4' Premandible present..6

5(4) Antenna weakly developed, 3rd segment not annulate. Procercus absent. BUCHONOMYIINAE...............*Buchonomyia* Fittkau

5' Antenna distinct, with segment 2 (or 2 and 3) often annulate. Procercus well developed. PODONOMINAE .... ***Parochlus*** Enderlein

6(4) Third antennal segment annulate. Prementum with three dense brushes of hair-like processes............................DIAMESINAE

6' Third antennal segment not annulate. Prementum with lamellae as opposed to brushes...7

7(6) Ventromental plates well developed, with setal beard beneath. Antenna 4-segmented................................PRODIAMESINAE

7' If ventromental plates developed, then either setal beard absent or antenna not 4-segmented...8

8(7) Prementum with brush-like appendage. Premandible with strong brush. TELMATOGETONINAE..9

8' Prementum without brush-like appendage. Premandible without brush, or if present, weak. ORTHOCLADIINAE...............102

9(8) Area anterior to frons with two distinct sclerites. S3 placed on a well-developed tubercle. Premandible with 3 rounded apical teeth... *Telmatogeton* Schiner

9' Area anterior to frons uniform, without delimited sclerites. S3 not on tubercle. Premandible simple........ *Thalassomyia* Schiner

10(1) Head capsule rounded to oval. Dorsomentum with teeth arranged in transverse plates or in longitudinal rows. Body with lateral dense fringe of setae 11

10' Head capsule more elongate. Dorsomentum indistinct, teeth reduced or absent. Lateral fringe of setae is absent, although isolated setae or groups of setae may be present. **Tribe Pentaneurini** 21

11 (10) Dorsomental teeth aligned in longitudinal rows, but not on distinct plates. Ligula with 6–7 pale teeth. **Tribe Clinotanypodini** 12

11' Dorsomental teeth arranged at margin of transverse or diagonal plates. Ligula with 4–5 pale or dark teeth 13

12(11) Ligula with 7 teeth (rarely 6 or 8); 1st inner tooth strongly bent toward outer tooth. Mandible slightly hooked. Dorsal anterior margin of body segment 4, with a pair of sclerotized hooks .... ***Coelotanypus*** Kieffer

12' Ligula with 6 teeth (rarely 5 or 7); 1st inner tooth slightly bent toward outer tooth. Mandible strongly hooked. Body segment 4 without sclerotized hooks .... ***Clinotanypus*** Kieffer

13(11) Mandible with enlarged base and short apical tooth. M appendage without pseudoradula. Pecten hypopharyngis absent. **Tribe Tanypodini** .... ***Tanypus*** Meigen

13' Mandible with slender base and longer apical tooth. M-appendage with pseudoradula. Pecten hypopharyngis present 14

14(13) Dorsomentum with continuous, concave-arched, toothed plate, weakly subdivided into median and lateral sections. Mandible with several rows of additional small dorsal and ventral teeth. **Tribe Fittkauimyiini** .... ***Fittkauimyia*** Karunakaran

14' Dorsomentum not continuous, toothed plate distinctly subdivided into median and lateral sections. Mandible without rows of additional teeth 15

15(14) Mandible with large rounded basal tooth. Ligula with 4 or 5 dark teeth; paraligula pectinate. **Tribe Procladiini** 16

15' Mandible with small basal tooth. Ligula with 5 dark or light teeth; paraligula bifid (rarely trifid). **Tribe Macropelopiini** 17

16(15) Antennal blade extending well beyond apex of flagellum. Ligula with 4 or 5 teeth .... ***Djalmabatista*** Fittkau

16' Antennal blade about as long as flagellum. Ligula with 5 teeth .... ***Procladius*** Skuse

17(15) Third antennal segment at least twice as long as wide. Ventrolateral mandibular setae all simple............. *Alotanypus* Roback
17' Third antennal segment shorter, about as long as wide. Ventrolateral mandibular setae 2 and 3 branched.................. 18

18(17) Apex of dorsomental plates with pointed medial extension, reaching almost to pseudoradula...............*Brundiniella* Roback
18' Apex of dorsomental plates not medially pointed, well separated from pseudoradula 19

19(20) Dorsomentum with 6–8 large inner teeth and 1 small tooth. Ligula with inner teeth directed forward........***Macropelopia*** Thienemann
19' Dorsomentum with 4 large inner teeth and 1 small tooth. Ligula with inner teeth curved outward...20

20(19) Ventral cephalic setae S10 simple ....*Apsectrotanypus* Fittkau
20' Ventral cephalic setae S10 branched..
....*Paggipelopia* Siri & Donato

21(10) Maxillary palp with 2 or more segments 22
21' Maxillary palp with single basal segment.............................23

22(21) Maxillary palp with 2 unequal segments. Pseudoradula broadened posteriorly......... *Zavrelimyia* (*Paramerina*) Fittkau in part
22' Maxillary palp with 2–6 segments; if only 2 segments, then segments are subequal in length. Pseudoradula not broadened posteriorly .... ***Ablabesmyia*** Johannsen

23(21) Posterior parapod with bifid or pectinate claws 24
23' Posterior parapod with simple claws.....................................27

24(23) Ligula with middle tooth longer than inner tooth. Posterior parapod with one pectinate claw or claws with only inner serrations. Anal tubules longer than parapods................***Nilotanypus*** Kieffer
24' Ligula with middle tooth distinctly longer than or subequal to inner tooth. Posterior parapod with one bifid claw. Anal tubules shorter than parapods...25

25(24) Ligula with middle tooth distinctly longer (sometimes subequal). Posterior parapod with a bifid claw with outer tooth shorter than inner tooth.......................................***Labrundinia*** Fittkau
25' Ligula with subequal teeth. Posterior parapod with bifid claws with outer tooth longer than inner tooth...............................26

26(25) Ligula with median and inner teeth lighter than outer teeth. Paraligula trifid......................................*Denopelopia* Roback

26' Ligula with all teeth equally dark. Paraligula bifid ....*Zavrelimyia* (*Zavrelimyia*) Fittkau

27(23) Posterior parapod with 1–2 pectinate claws, dark or light, with inner margin serrated.............................. *Monopelopia* Fittkau

27' Posterior parapod without serrated claws..............................28

28(27) Ring organ of maxillary palp, situated near apex of basal segment. Mandible with no basal tooth. *Thienemannimyia* group 29

28' Ring organ of maxillary palp, situated usually near middle of basal segment. Mandible with weakly to moderately developed basal tooth...30

29(28) Distance between ventrolateral setae 2 and 3 of mandible ½ as great as that between seta 2 and sensillum minusculum ....*Metapelopia* Silva, Oliveira & Trivinho-Strixino

29' Distance between ventrolateral setae 2 and 3 of mandible smaller than ½ distance between seta 2 and sensillum minusculum...*Thienemannimyia* Fittkau

30(28) Ligula with median and inner teeth shorter than outer teeth. Supraanal not arising from dark sclerotized area. Anal tubules shorter...***Larsia*** Fittkau

30' Ligula with subequal teeth. Supraanal seta arising from dark sclerotized area. Anal tubules are long and thin, surpassing length of posterior parapods 31

31(30) Head usually dark brown. Posterior parapods with dark claws .... *Hudsonimyia* Roback

31' Head pale-yellow. Posterior parapods with light claws...........32

32(31) Cephalic seta S10 between S9 and ventral pore (VP), forming an 80°–90° angle...***Pentaneura*** Philippi

32' Ventral pore (VP) between cephalic setae S9 and S10, forming a straight-line diagonal to longitudinal axis of head capsule .... *Parapentaneura* Stur, Fittkau & Serrano

33 (2) Antenna mounted on distinct elongated base; Lauterborn organs often mounted on short to long pedicels. **Tribe Tanytarsini**......34

33' Antenna not mounted on distinct elongated base; Lauterborn organs not mounted on pedicels..47

34(33) Ventromental plates narrow, separated medially by at least the width of the 3 median teeth. Lauterborn organs not mounted on long pedicels. Larvae build transportable cases 35
34' Ventromental plates wide, almost touching each other medially. Lauterborn organs mounted on long pedicels. If larvae in cases, cases not transportable 38

35(36) One of the Lauterborn organs at the base of antennal segment 2 and the other placed apically 36
35' Both Lauterborn organs at apex of antennal segment 2 37

36(35) Premandible with 4 teeth .... *Zavrelia* Kieffer
36' Premandible with 2–3 teeth .... *Stempellinella* Brundin

37(35) Antennal base with palmate process. Frontoclypeal setae simple or bifid .... ***Stempellina*** Thienemann & Bause
37' Antennal base with simple spur. Frontoclypeal setae branched .... *Constempellina* Brundin/Tanytarsini genus A Ekrem

38(34) Premandible with 3–5 main teeth 39
38' Premandible with 2 main teeth 43

39(38) Lauterborn organs small, sessile. Larvae living in marine habitats .... *Pontomyia* Edwards
39' Lauterborn organs on distinct pedicels. Larvae living in freshwater 40

40(39) Antennal segment 2 as long as or shorter than antennal segment 3; Lauterborn organs large, almost as large as their pedicels .... ***Cladotanytarsus*** Kieffer
40' Antennal segment 2 longer than segment 3; Lauterborn organs small, on long pedicels 41

41(40) Pedicels of Lauterborn organs usually with basal half, more sclerotized or annulated .... ***Tanytarsus ortoni* group**
41' Pedicels of Lauterborn organs usually less sclerotized .... ***Tanytarsus*** v. d. Wulp

42(38) Pecten epipharyngis one plate with 3–5 lobes 43
42' Pecten epipharyngis one or three plates with several apical teeth 44

43(42) Lauterborn organs on pedicels longer than flagellum. Mandible with distinct outer hump .... *Sublettea* Roback

43' Lauterborn organs sessile or on pedicels shoter than flagellum. Mandible without outer hump .... ***Paratanytarsus*** Thienemann & Bause in part

44(42) Lauterborn organs on short pedicels, not surpassing the flagellum 45

44' Lauterborn organs on longer pedicels 46

45(44) Mentum with 4 or 5 pairs of lateral teeth. Lauterborn organs on moderately long pedicels .... *Neozavrelia* Goetghebuer

45' Mentum with 5 pairs of lateral teeth. Lauterborn organs on short pedicels .... ***Rheotanytarsus*** Thienemann & Bause

46(44) Pedicels 3–5 times longer than combined length of segments 3–5 .... ***Micropsectra*** Kieffer

46' Pedicels slightly longer than combined length of segments 3–5 .... ***Paratanytarsus*** in part

47(33) SI and SII simple; SII often large and blade like. Pecten epipharyngis rounded or subtriangular, consisting of a single plate or scale, which may be simple, serrated, notched, or toothed. Mandible without dorsal tooth. Maxillary palps often very long, as long as half of the 1st antennal segment. **Tribe Chironomini** in part 48

47' SI plumose or fringed, SII never blade like. Pecten epipharyngis a wide multitoothed comb or three separate small plates, toothed apically (sometimes smooth). Mandible usually with dorsal tooth. Maxillary palps usually reduced 59

48(47) Mentum concave with one large pale tooth, or with a median large gap between the dark lateral teeth 49

48' Mentum convex or linear 51

49(48) Antenna 7-segmented. Mentum with 7 pairs of lateral teeth .... *Demicryptochironomus* Lenz

49' Antenna 5–6-segmented. Mentum with 4–5 pairs of lateral teeth 50

50(49) SI minute, less than ⅕ length of SII. Premandible without brush. Lateral teeth of mentum obliquely arranged ...... *Gillotia* Kieffer

50' SI at least ½ as long as SII. Premandible with brush. Lateral teeth of mentum not obliquely arranged ...... ***Cryptochironomus*** Kieffer

51(48) Antenna 5-segmented 52
51' Antenna 6–7-segmented 57

52(51) Anterior margin of ventromental plate serrated or wavy .... ***Parachironomus*** Lenz
52' Anterior margin of ventromental plate smooth 53

53(52) Premandible apically bifid 54
53' Premandible with 3 or more teeth 56

54(53) Antennal blade as long or longer than flagellum. Mentum with median tooth trifid .... *Microchironomus* Kieffer
54' Antennal blade shorter than flagellum. Mentum with single median tooth (occasionally notched appearing trifid) 55

55(54) First antennal segment short, 2–2.5 times longer than wide. Median tooth broadly rounded or laterally notched, extending far forward of lateral teeth........... *Cryptotendipes* Beck & Beck
55' First antennal segment longer than above. Median tooth broad notched, may be double, not extending far forward of lateral teeth .... ***Cladopelma*** Kieffer

56(53) Antennal segments 2 and 3 subequal. Ventromental plates weakly striated .... *Harnischia* Kieffer
56' Antennal segment 2 much longer than segment 3. Ventromental plates coarsely striated .... *Paracladopelma* Harnisch

57(51) Antenna 7-segmented. Mandible with modified inner teeth .... *Robackia* Reiss
57' Antenna 6-segmented. Mandible without modified inner teeth 58

58(57) Premandible with 3 large inner teeth; brush hyaline and not easily observed .... *Saetheria* Jackson
58' Premandible with 2 large apical teeth and usually 1 smaller proximal tooth; brush dense and conspicuous .... ***Pelomus*** Reiss

59(47) Ventromental plates bar-like, in near contact medially. Seta subdentalis dorsal on the same side of mandible as seta interna. **Tribe Pseudochironomini** 60
59' Ventromental plates variable in shape. Seta subdentalis ventral, on the opposite side of mandible from seta interna (rarely absent). **Tribe Chironomini** in part 63

60(59) Pecten mandibularis present. Inner mandibular teeth and mentum yellow pale.. *Aedokritus* Roback
60' Pecten mandibularis weak or absent. Inner mandibular teeth and mentum darkened..61

61(60) SI arising from separate bases....... ***Pseudochironomus*** Malloch
61' SI setae arising from a common base....................................62

62(61) Mentum with 2nd lateral tooth minute, fused to 1st lateral tooth .... *Manoa* Fittkau
62' Mentum with 2nd lateral tooth distinct, never fused to 1st lateral... *Riethia* Kieffer

63(59) Mentum with median tooth or teeth deeply recessed.............64
63' Mentum with median tooth or teeth not deeply recessed.......65

64(63) Antenna 5-segmented. Mandible with slender seta subdentalis, seta interna and pecten mandibularis present. Mentum with median tooth broad and extended.............. *Hyporhygma* Reiss
64' Antenna 6-segmented. Mandible with enlarged seta subdentalis, seta interna and pecten mandibularis absent. Mentum with median tooth deeply divided....***Fissimentum*** Cranston & Nolte

65(63) Mentum concave with 8–12 teeth. Ventromental plates without well-defined striae. Larvae mining in dead submerged wood or in submerged leaves...66
65' Mentum variable. Ventromental plates usually with numerous striae. Larvae in a variety of habitats....................................67

66(65) Mentum with 8–10 teeth.....***Xestochironomus*** Sublette & Wirth
66' Mentum with 10–12 teeth................ ***Stenochironomus*** Kieffer

67(65) Larva with body segment 11, with dorsal hump....................68
67' Larva with body segment 11 without dorsal hump................69

68(67) Seta submenti simple. Hump on segment 11 anteriorly directed ....***Zavreliella*** Kieffer
68' Seta submenti plumose. Hump on segment 11 posteriorly directed........................ ***Lauterborniella*** Thienemann & Bause

69(67) Larva with one or two pairs of ventral tubules on the 8th abdominal segment..70
69' Larva without ventral tubules on the 8th abdominal segment 76

70(69) Mandible with a row of striae spreading radially near base 71
70' Mandible without a row of striae spreading radially near base 72

71(70) Mentum with median tooth projecting far beyond lateral teeth and 5th lateral tooth larger than 4th or 6th. Pecten epipharyngis consists of 3 scales, bearing minute spinules. Eighth abdominal segment with 1 pair of ventral tubules.............*Benthalia* Lipina
71' Median tooth of mentum not projecting far beyond lateral teeth. Pecten epipharyngis has a broad multitoothed comb. Eighth abdominal segment with 1 or 2 pairs of ventral tubules.. ***Chironomus*** Meigen in part

72(70) Premandible with at least 5 apical teeth .... *Kiefferulus* Goetghebuer in part
72' Premandible apically bifid...73

73(72) Seta subdentalis with ventral margin fringed or toothed. Ventromental plates curved medially, almost touching each other..................................***Goeldichironomus*** Fittkau in part
73' Seta subdentalis simple. Ventromental plates not curved and medially separated..74

74(73) Width of ventromental plate smaller than width of mentum. Pecten epipharyngis with less than 10 rounded lobes.. ***Dicrotendipes*** Kieffer in part
74' Width of ventromental plate larger than or subequal to width of mentum. Pecten epipharyngis consists of a comb with at least 10 or more teeth or faintly tripartite and bearing minute spinules 75

75(74) Dorsum of head capsule with 1 medial labral sclerite, anterior to frontoclypeal apotome; apotome with large fenestra ....*Einfeldia* Kieffer
75' Dorsum of head capsule with 2 medial labral sclerites, anterior to frons; apotome without fenestra .... *Glyptotendipes* Kieffer in part

76(69) Mentum with an even number of teeth...................................77
76' Mentum with an odd number of teeth....................................92

77(76) Antenna 5-segmented..78
77' Antenna 6-segmented..84

78(77) Mentum with median teeth higher or subequal to the 1st lateral teeth..79
78' Mentum with median teeth lower than the 1st lateral teeth 81

79(78) Mentum with median teeth higher than the 1st lateral teeth ....***Polypedilum*** in part
79' Mentum with all teeth subequal..80

80(78) Ventromental plates elongate with anterior margin smooth ....***Polypedilum*** (***Asheum***) Sublette & Sublette
80' Ventromental plates shorter with anterior margin wavy ....***Polypedilum*** Kieffer in part

81(78) Frons separated from clypeus..82
81' Frontoclypeal apotome present..83

82(81) Mandible with notch at base of inner teeth. Mentum strongly arched, with 11–13 teeth...........***Endotribelos*** Grodhaus in part
82' Mandible without notch. Mentum not strongly arched, with 15 teeth..***Endochironomus*** Kieffer

83(81) Anterior margin of frontoclypeal apotome convex, with 1 medial labral sclerite anterior to it .... *Phaenopsectra* Kieffer
83' Anterior margin of frons straight, with 2 medial labral sclerites anterior to it..*Tribelos* Townes

84(77) Ventromental plate with each stria ending in a spine just posterior to plate margin.............................. *Imparipecten* Freeman
84' Ventromental plate without spines..85

85(84) Mentum with 2 median teeth..86
85' Mentum with 4 median teeth..89

86(85) Mentum with all teeth even colored, bearing 7 pairs of lateral teeth; 3rd lateral tooth larger than 2nd lateral tooth ....*Sigmoitendipes* Andersen, Mendes & Pinho
86' Mentum with median teeth pale, bearing 6 pairs of lateral teeth; 3rd lateral tooth smaller than 2nd lateral tooth 87

87(86) Mentum with 2 recessed, small median teeth; 2nd lateral teeth higher than median teeth, fused to 1st lateral tooth. Ventromental plates trapezoid............ ***Oukuriella*** Epler in part

87' Mentum with 2–3 pale median teeth that are subequal to 2nd lateral teeth. Ventromental plates not trapezoid.................... 88

88(87) Frontoclypeal apotome present. Mentum with 2 pale median teeth. Maxillary plate conspicuous .... ***Apedilum*** Townes

88' Frons separated from clypeus. Mentum with 3 pale median teeth (central median tooth may be minute). Maxillary plate inconspicuous.................................... ***Microtendipes*** Kieffer

89(85) Pecten epipharyngis undivided...
.... *Claudiotendipes* Andersen, Mendes & Pinho

89' Pecten epipharyngis divided in three plates........................... 90

90(89) Mentum with at least the outer pair higher than the remaining lateral teeth....................................... *Stictochironomus* Kieffer

90' Mentum with median teeth lower than or subequal to 2nd lateral teeth; 2nd outer pair higher than remaining lateral teeth... 91

91(90) SI plumose with separate bases. Mentum with median teeth darkened; central pair of the 4 median teeth lower and more slender than outer median teeth...................... *Omisus* Townes

91' SI setae with bases fused. Mentum with median teeth pale; central pair of the 4 median teeth equal to or higher than outer median teeth.. ***Paratendipes*** Kieffer

92(76) Mentum with median tooth mostly pale............................... 93

92' Mentum with all teeth evenly colored.................................. 95

93(92) Frons strongly narrowed anteriorly. Mandible without dorsal tooth. Mentum with median tooth, broad and higher than 1st lateral teeth....................................... ***Paralauterborniella*** Lenz

93' Frons not narrowed anteriorly. Mandible with 1–2 dorsal teeth. Mentum with median tooth shorter than 1st lateral teeth
.... 94

94(93) Mandible with double dorsal tooth, seta subdentalis enlarged. Pecten epipharyngis in three distally toothed plates..................
.... ***Beardius*** Reiss & Sublette

94' Mandible with simple dorsal tooth, seta subdentalis slender. Pecten epipharyngis in three smooth plates...............................
.... ***Oukuriella*** Epler in part

95(92) Head capsule slender, rectangular. Mandible set very far forward.. *Kribiodorum* Kieffer
95' Head capsule more rounded. Mandible not set far forward 96

96(95) Mandible with apical tooth very slender and long; dorsal tooth absent. Small larvae.................................. ***Nilothauma*** Kieffer
96' Mandible with apical tooth shorter; dorsal tooth present. Usually larger larvae...97

97(96) Labral sclerite 1 absent...98
97' Labral sclerite 1 present.. 99

98(97) Frons separated from clypeus. Mentum with one median simple tooth, sometimes slightly notched......... ***Endotribelos*** Grodhaus in part
98' Frons fused to clypeus. Mentum with median tooth trifid .... ***Chironomus*** Meigen in part

99(97) Labrum with large anterolateral setal brushes on each side. Larvae mining in sponges.................. ***Xenochironomus*** Kieffer
99' Labrum without anterolateral setal brushes. Larvae in a variety of habitats 100

100(99) Ventromental plates well separated medially .. ***Dicrotendipes*** Kieffer in part
100' Ventromental plates almost touching each other medially . 101

101(100) Seta subdentalis on mandible with ventral margin fringed. Ventromental plates strongly curved posteromedially .... ***Goeldichironomus*** Fittkau in part
101' Seta subdentalis simple. Ventromental plates not curved .... ***Axarus*** Roback

102(8) Anal end without procercus or, if present, without distinct anal setae...103
102' Anal end with procercus present, with variable number of anal setae...112

103(102) Scales of pecten epipharyngis divided into 2–3 teeth, forming continuous row of about 8 teeth. SI palmate..... *Antillocladius* Sæther
103' Pecten epipharyngis consists of 3 scales. SI palmate only in *Smittia*...104

104(105') Preanal and anal segments curved at right angles to axis of rest of body..105
104' Preanal and anal segments in same axis as rest of body.....106

105(104) Posterior parapods divided into 2 parts .... ***Gymnometriocnemus*** Edwards
105' Posterior parapods undivided .... ***Bryophaenocladius*** Thienemann

106(104) Anal tubules absent. Mainly marine................................107
106' Anal tubules present. Never marine................................108

107(106) SI and SII broad and apically feathered.........*Clunio* Haliday
107' SI feathered, SII always simple... .... *Thalassosmittia* Strenzke & Remmert

108(106) Mandible with spine-like teeth........... *Symbiocladius* Kieffer
108' Mandibular with rounded or subtriangular teeth............109

109(108) SI and SII bifid..110
109' SI bifid or not, SII never bifid..111

110(109) Posterior parapods with 0–5 short claws. Antennal blade extending beyond flagellum...............*Pseudosmittia* Edwards
110' Posterior parapods with 7–12 long claws. Antennal blade not, or only slightly, extending beyond flagellum........*Allocladius* Kieffer

111(109) SI simple. Mandible without seta interna........... *Mesosmittia* Brundin
111' SI plumose. Mandible with seta interna........ *Smittia* Holmgren

112(102) Mentum with equal sized median and 5 pairs of large, lateral teeth. Larvae phoretic on catfish ....*Ichthyocladius* Fittkau
112' Mentum variable in shape...113

113(112) Antenna at least half-length of head 114
113' Antenna shorter than half-length of head 118

114(113) Antenna with whip-like or thread-like terminal segment 115
114' Antenna with short terminal segment, but it may have short hair-like extension 116

115(114) SI simple. Mandible with 3–4 inner teeth..........*Stictocladius* Edwards
115' SI bifid. Mandible with 5 inner teeth......*Lopescladius* Oliveira

116(114) Antenna 4-segmented, generally longer than head...***Corynoneura*** Winnertz
116' Antenna 5-segmented, usually shorter than head 117

117(116) Sub-basal seta of posterior parapod simple...***Thienemanniella*** Kieffer
117' Sub-basal seta of posterior parapod split..*Ubatubaneura* Wiedenbrug & Trivinho-Strixino

118(113) Mentum with at least 16 teeth...***Orthocladius*** van der Wulp in part
118' Mentum with no more than 15 teeth.............................119

119(118) Ventromental plates with beard beneath..........................120
119' Ventromental plates without beard beneath....................126

120(119) Antenna 6-segmented; 6th segment small, hair-like. Premandible without brush...***Parakiefferiella*** Thienemann in part
120' Antenna 4- or 5-segmented. Premandible with brush.......121

121(120) Antenna 4-segmented..................***Orthocladius*** van der Wulp .... in part
121' Antenna 5-segmented 122

122(120) Abdomen with alternating simple and plumose lateral setae ....***Synorthocladius*** Thienemann
122' Abdomen without alternating simple and plumose setae, but single pairs of setal tufts may be present posterolaterally 123

123(122) Mandible with 4 inner teeth, apical tooth short ....*Diplocladius* Kieffer
123' Mandible with 3 or fewer inner teeth or when with 4, apical tooth longer than combined width of 4 inner teeth or anal tubules absent...124

124(123) SI broadly trifid, palmate, or with 4 long, narrow teeth .... ***Psectrocladius*** Kieffer in part
124' SI usually bifid; occasionaly simple, apically split into 4 or more short teeth, plumose or coarsely serrate..................125

125(124) Apical tooth of mandible longer than combined width of 3 inner teeth. Ventromental plates large and triangular. Head capsule without ventral tubercle.. .... *Psectrocladius* Kieffer in part

125' Apical tooth of mandible shorter than or subequal to combined width of 3 inner teeth. If ventromental plates large and triangular, then head capsule with pair of ventral tubercles .... *Rheocricotopus* Brundin

126(119) Ventromental plates well developed, extending beyond lateral margin of mentum.. 127

126' Ventromental plates absent or vestigial or if present not extending well beyond lateral margin of mentum or plates very thin.. 131

127(126) All S setae simple. Mentum with small pair of median teeth, often well separated from the 0–6 pairs of lateral teeth, which may be small and fused or closely appressed to each other; ventromental plates long and elongate........................ .... *Nanocladius* Kieffer

127' SI never simple. Mentum variable in shape; ventromental plate broader and less elongate....................................... 128

128(127) Antenna 7-segmented; 7th segment hair-like; 3rd segment minute, ⅓ length of 4th........................ *Heterotrissocladius* Spärck

128' Antenna 5–6 segmented; 3rd segment never as short as ⅓ the length of 4th.. 129

129 (128) Ventromental plates single. Seta submenti near posterior margin of ventromental plates or more posteriad................. .... *Parakiefferiella* Thienemann in part

129' Ventromental plates double. Seta submenti well anterior to posterior margin of ventromental plates.......................... 130

130(129) Antenna with long basal segment, antennal ratio > 1.25. Antennal blade shorter than flagellum.................................. .... *Parametriocnemus* Goetghebuer

130 Antenna with shorter basal segment, antennal ratio 0.5–1.0. Antennal blade shorter or longer than flagellum................... .... *Paraphaenocladius* Thienemann

131(126) Antenna 6–7-segmented; terminal segment vestigial, hair-like. Ventromental plates moderately developed.. *Parakiefferiella* Thienemann in part

131' Antenna with 5 or fewer segments; last segment not hair-like. Ventromental plates moderately developed to absent 132

132(131) Abdomen with long simple setae, at least half as long as the segment holding them.. 133

132' Abdomen without long simple setae, or if long setae present, they are arranged as pairs of setal tufts.......................... 134

133(132) SI bifid, with several apical teeth or plumose.. ***Metriocnemus*** van der Wulp in part

133' SI simple without any apical dentations.. ***Eukiefferiella*** Thienemann in part

134(132) Mentum with 3 median teeth. SII strong, situated on tubercle 135

134' Mentum with 1–2 median teeth. SII not on tubercle........ 136

135(134) Antenna about ¼ as long as head capsule. Anterior parapods with two well sclerotized, large hook-like claws... *Tempisquitoneura* Epler

135' Antenna about ⅓ as long as head capsule. Anterior parapods without sclerotized claws....... *Onconeura* Andersen & Sæther

136(134) SI simple.. 137

136' SI bifid, pectinate, plumose, and serrate........................ 143

137(136) Inner margin of mandible with spines........................... 138

137' Inner margin of mandible smooth................................ 141

138(137) Procercus reduced, with 2 setae thicker than the others on each procercus .... *Cardiocladius* Kieffer in part

138' Procercus at least as long as wide; setae about equally thick 139

139(138) Mentum with 4 pairs of lateral teeth.. ***Eukiefferiella*** Thienemann in part

139' Mentum with 5 pairs of lateral teeth............................. 140

140(139) Anal setae much shorter than posterior parapods .... *Cardiocladius* Kieffer in part
140' Anal setae longer than posterior parapods............................ ....***Eukiefferiella*** Thienemann in part

141(137) Mentum with large bifid median tooth and only 2 lateral teeth..*Austrobrillia* Freeman
141' Mentum with single or moderately developed bifid median tooth and at least 5 pairs of lateral teeth......................... 142

142(141) Body with posterolateral setal tufts. Mentum with single median tooth..................... ***Cricotopus*** van der Wulp in part
142' Body without posterolateral setal tufts. Mentum with bifid median tooth...............................***Limnophyes*** Eaton in part

143(136) Labral lamellae well developed..144
143' Labral lamellae absent or vestigial..................................146

144(143) Mentum with 2 elongate median teeth (a small central tooth can be present between median teeth); setae submenti displaced posteriorly... ***Brillia*** Kieffer
144' Mentum with 2–4 median teeth, none elongate; setae submenti displaced anteriorly, near base of mentum............. 145

145(144) Procerci well developed, at least twice as long as wide. Supraanal setae shorter than anal tubules ....***Metriocnemus*** van der Wulp in part
145' Procerci weakly developed, about as long as wide. Supraanal setae as long as, or longer than, anal tubules........................ .... *Thienemannia* Kieffer

146(143) Antennal blade longer than flagellum. SI apically pectinate or plumose.. 147
146' Antennal blade subequal in length to flagellum. SI serrate or palmate... 149

147(146) Mentum with two median teeth ...... *Phytotelmatocladius* Epler
147' Mentum with single median tooth..................................148

148(147) Abdominal segments with one pair of setal tufts. Fourth antennal segment as long as the 3rd ....*Gravatamberus* Mendes & Andersen
148' Abdominal segments without setal tufts. Fourth antennal segment about twice as long as the 3rd................................ ....*Gynocladius* Mendes, Sæther & Andrade-Morraye

149(146) SI palmate. Seta subdentalis subovoid curved........................ ....*Parapsectrocladius* Cranston

149' SI serrate. Seta subdentalis narrow and simple................150

150(149) Mandible with 3 inner teeth. Supraanal setae about as long as anal setae.................................... ***Limnophyes*** Eaton in part

150' Mandible with 4 inner teeth. Supraanal setae about ⅓ as long as anal setae.................................... *Compterosmittia* Sæther

TERMINOLOGY AND MORPHOLOGY

There is a substantial difference between the morphology of a living larva and its subfossil remains. Subfossil larvae preserved in sediments are typically without a body; only the chitinized head capsule remains, which is incomplete (for comparison of a head of living and subfossil larva, see Figure 3.1). Delicate, but for identification, essential features, such as labral setae SI and SII, labral lamellae and pecten epipharyngis, are practically always missing; more robust structures, such as premandible and antenna, are also rarely present. What is usually left is the mentum and mandibles, that is, these are the features we have to rely upon for identification. Often these structures are not associated with the head capsule either. The degree of disarticulation of the remains depends on many taphonomic factors (e.g., hydromorphology of the lake, chemical processes and type of sediment), but also on the features of the subfamily where the remains belong. Head capsules of Orthocladiinae and some Chironomini tend to split in half while heads of Tanytarsini and Tanypodinae are basically

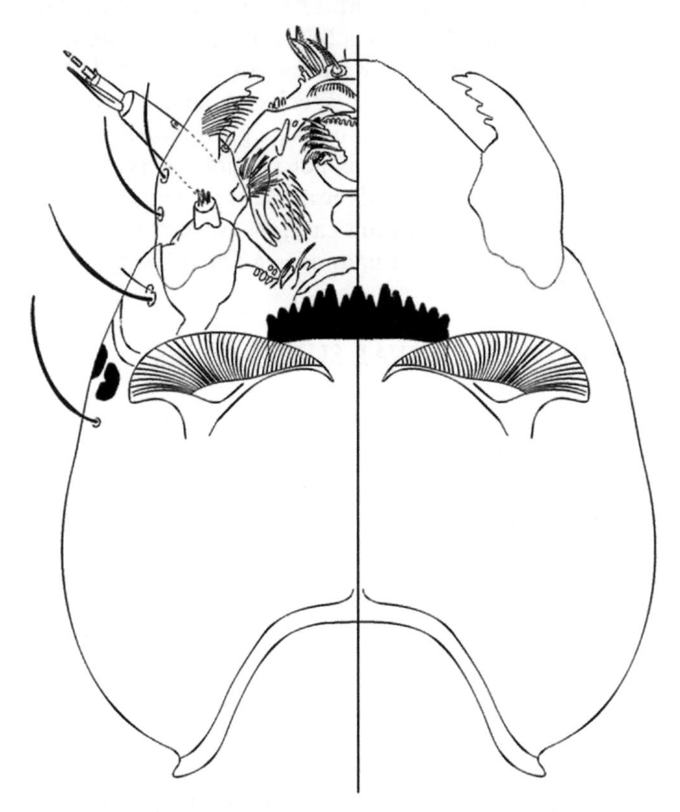

Figure 3.1 Comparison of morphological features of the head of a living larva (left) and a common subfossil remnant (right).

DOI: 10.1201/9780429021572-3

always entire; subfossil remains of the latter are often torn and miss most of the structures within the head capsule (Brooks et al. 2007).

All in all, when identifying subfossil remains, we are relying on limited number of characters, such as number of teeth and shape of mentum, ligula and mandible, shape of ventromental plates, size of antennal pedestal, presence/absence of a spur on it, etc. (for detailed morphology of subfossil remains, see Figures. 4.1, 4.29, 5.1, and 7.1). However, in well-preserved specimens, additional structures may be visible, and were used in order to confirm the accuracy of identification or to distinguish morphotypes. Morphology of individual subfamilies is shown in the beginning of each chapter dedicated to the subfamily. The description of genera follows Brooks et al. (2007), Cranston (2010), and Andersen et al. (2013), unless stated otherwise.

For detailed description of the morphology of subfossil chironomids, along with field and laboratory methods, see Brooks et al. (2007). Comprehensive morphology of living larvae can be found in Epler (2001), Cranston (2013), and Silva et al. (2018).

In this book, besides presenting morphotypes previous described, there are also several new records for Central America. The names of the known morphotypes follow Brooks et al. (2007), e.g., *Chironomus plumosus*-type, while for the new ones, new names were designated. Therefore, if the type could be associated with a species (or species group), the word "-type" was used after the species name, e.g., *Endotribelos hesperium*-type. Several times, however, it was not possible to associate the larval remain with an existing species and the name of the lake of origin was used to designate the morphotype. In this case, the word "type" before the site name was our chosen option for naming, e.g., *Beardius* type Chanmico.

KEY TO SUBFOSSIL LARVAL SUBFAMILIES

1 Mentum and ventromental plates absent; distinctive structure (ligula) usually with 5 teeth present. Antenna long, capable of complete retraction into head...TANYPODINAE

1' Mentum strongly developed, with or without ventromental plates. Antenna not retractile..2

2 Mentum expanded laterally to ventromental plates that are usually striated and always without beard..3

2' Mentum without ventromental plates, or, if plates present, they are never striated (some *Nanocladius* may have a few markings on ventromental plates) and usually with beard.......................................5

3 Mentum with large, fan-shaped ventromental plates with striations, widely separated medially, antennal pedestal small, without large spur. CHIRONOMINAE in part........................Tribe Chironomini

3' Mentum with narrow ventromental plates with striations, usually close to each other medially, antennal pedestal prominent, with or without spur, antenna, if present, usually very long 4

4' Antenna mounted on distinct pedestals, Lauterborn organs usually well developed and often situated on short to long pedicels. CHIRONOMINAE in part .... Tribe Tanytarsini*

4' Antenna not mounted on distinct pedestals, Lauterborn organs not placed on pedicels. CHIRONOMINAE in part .... Tribe Pseudochironomini*

5 Premandible always present; mentum usually with 4–6 lateral teeth, mandible usually with 3–4 teeth ...................ORTHOCLADIINAE

5' Premandible, ventromental plates and beard absent; mentum with 7 lateral teeth; mandible with 5 and more teeth .... PODONOMINAE

Keys to genera of subfossil subfamilies are ordered alphabetically and can be found on the following chapters:

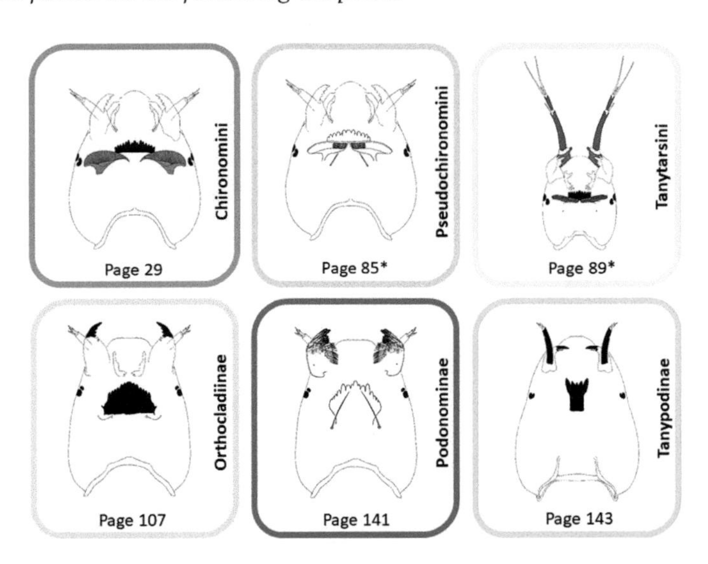

* The distinction between tribes Pseudochironomini and Tanytarsini is mainly based on characteristics of the antenna, which is a feature that may be missing on subfossil remains. Therefore, in case of absence of this trait, the material must be keyed considering the keys for both tribes in the next chapter.

TRIBE CHIRONOMINI

The typical features of the Chironomini tribe are the following: ventromental plates vary in shape but are usually well developed and striated (this feature is secondarily reduced in the *Stenochironomus* complex). Antenna not placed on distinct pedestals; Lauterborn organs not situated on pedicels. Seta subdentalis located ventrally, on the opposite side of mandible from seta interna (Cranston et al. 2012). Figure 4.1 shows the Chironomini larval head capsule with common subfossil features.

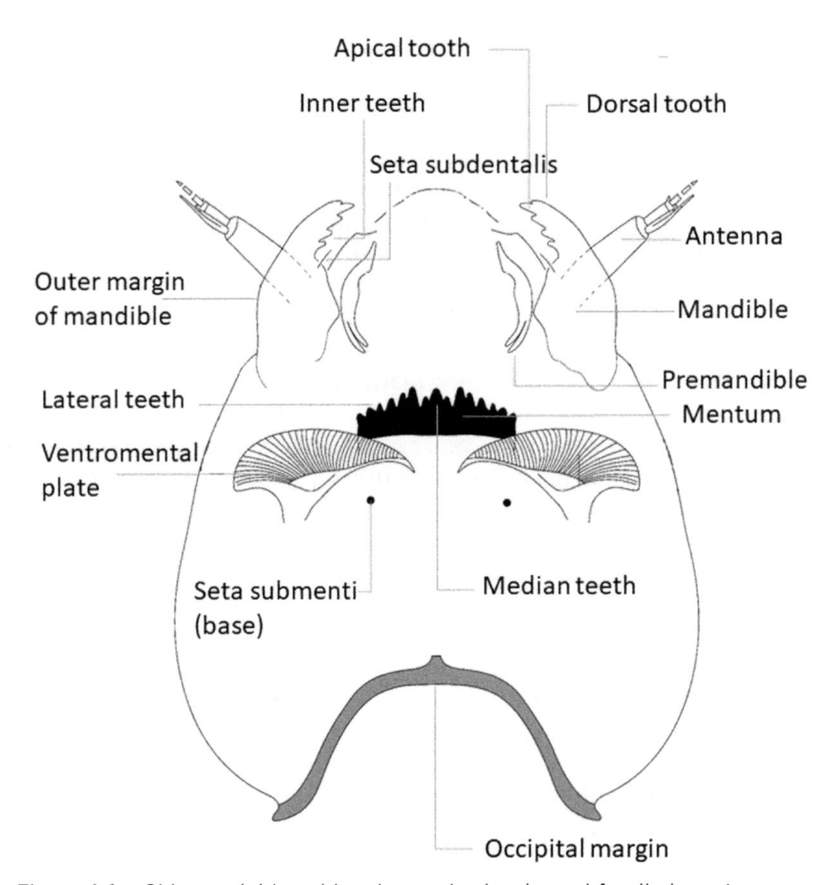

Figure 4.1 Chironomini larval head capsule showing subfossil characters.

DOI: 10.1201/9780429021572-4

KEY TO GENERA OF CHIRONOMINI*

1 Ventromental plates inconspicuous, striae absent, mentum strongly concave with distinctive pattern of teeth 2

1' Ventromental plates distinct, usually with numerous striae, mentum variable..3

2(1) Mentum with 8 teeth...*Xestochironomus*

2' Mentum with 10 teeth ....*Stenochironomus*

3(1) Mentum with median tooth single, without central apical notch 4

3' Mentum with 2 or more median teeth, or, if single, with central apical notch..11

4(3) Mentum with median tooth much shorter than 1st lateral, which may be notched, and 5–6 pairs of lateral teeth 5

4' Mentum with median tooth subequal or higher than 1st lateral, and 6 or more pairs of lateral teeth...7

5(4) Ventromental plates wide and strongly curved medially, separated by about the width of a median tooth; seta subdentalis large with toothed inner margin, triangulum occipitale well developed, mandible strongly tuberculate. .... *Goeldichironomus* in part

5' Ventromental plates narrow, fan-like, 1st lateral with a notch on inner margin (not visible in worn down specimens) 6

6(5) Mentum with median with median tooth lighter than remainder of mentum, 2 inner teeth on mandible, lateral end of ventromental plates pointing backward or laterally......................*Beardius*

6' Mentum with median tooth the same color as mentum, 3 inner teeth on mandible, lateral end of ventromental plates pointing forward.. *Oukuriella*

7(4) Mentum with 6 pairs of lateral teeth.......................................8

7' Mentum with 7 pairs of lateral teeth....................................10

8(7) Mentum with median tooth narrow (up to 2× as broad as 1st lateral tooth), strongly pigmented, ventromental plates narrow (about half-width of mentum), fan shaped .... *Dicrotendipes*

8' Mentum with median tooth large, domed, and pale, ventromental plates wider than mentum...9

9(8) Mentum concave, outermost lateral tooth notched apically .... *Cryptochironomus*
9' Mentum convex, outermost lateral tooth not notched .... *Paralauterborniella*

10(7) Mentum with median tooth broad domed, darkly pigmented, flanked by minute accessory teeth, 2nd lateral tooth shorter than 1st and 3rd lateral teeth .... *Xenochironomus*
10' Mentum with median tooth narrow and pointy, anterior margin of ventromental plates scalloped .... *Parachironomus*

11(3) Mentum with 2 median teeth.. 12
11' Mentum with 3 or more median teeth.................................... 20

12(11) Mentum with 5–6 lateral teeth... 13
12' Mentum with 7 lateral teeth (7th may be minute).................. 17

13(12) Mentum with median teeth pale, slightly higher than 1st lateral, minute accessory teeth may be present between median teeth and 1st lateral, in worn down specimens not visible or appearing as a notch on 1st lateral .... *Apedilum*
13' Mentum uniformly colored, median teeth not paler than lateral teeth 14

14(13) Mentum with 1st lateral tooth minute, shorter than median and 2nd lateral tooth.. 15
14' Mentum with 1st lateral tooth subequal to median and 2nd lateral tooth, mentum rather horizontal in appearance, mandible characteristic: very long, slender apical tooth, 4 inner teeth on a different focal plane, increasing in size from 1st to 4th, semicircular in appearance... *Nilothauma*

15(14) Ventromental plates strongly curved; mentum divided by a line connecting the inner apical margin of ventromental plates and the base of 2nd lateral tooth (usually visible). .... *Microtendipes* in part
15' Ventromental plates wedge-shaped with straight apical margin, mentum not divided by a line linking ventromental plates and base of 2nd lateral tooth 16

16(15) Mentum with 1st and 2nd lateral teeth separate, seta submenti branched .... *Lauterborniella*
16' Mentum with 1st and 2nd lateral teeth partially fused, seta submenti simple.. *Zavreliella*

17(12) Ventromental plates very broad, mentum with deeply sunken pair of median teeth..*Fissimentum*

17' Ventromental plates shorter, mentum with median teeth subequal or higher than 1st lateral..18

18(17) Mentum with outermost 3 lateral teeth arranged in distinct cluster, 6th lateral tooth higher than adjacent teeth..... *Cladopelma*

18' Mentum with outermost 3 lateral teeth not arranged in distinct cluster and subequal in size..19

19(18) Mentum distinctive, with 1st lateral tooth much lower than median and 2nd lateral tooth..................... *Polypedilum* in part

19' Mentum with teeth subequal in size........... *Polypedilum* in part

20(11) Mentum with 3 median teeth (outer median teeth may be minute and partly fused with central tooth)................................21

20' Mentum with 4 median teeth..26

21(20) Mentum with 5 pairs of lateral teeth, regularly decreasing in size, outermost minute; mandible strongly curved with very long apical tooth, 3 (4?) inner teeth diminishing in size toward base, dorsal tooth absent........................... *Harnischia*-complex

21' Mentum with 6 or more pairs of lateral teeth, mandible of different shape...22

22(21) Mentum with 1st lateral tooth shorter than outer median and 2nd lateral teeth..23

22' Mentum with 1st lateral tooth subequal to outer median and 2nd lateral teeth or higher..24

23(22) Mentum strongly arched, all teeth uniformly colored, central median tooth may be apically notched, 6th lateral may be reduced/missing; mandible with large incised area at the base of the inner teeth; ventromental plates not coarsely striated .... *Endotribelos* in part

23' Mentum rather horizontal, 3 median teeth pale (median tooth may be minute); mandible without large incised area; ventromental plates coarsely striated...................*Microtendipes* in part

24(22) Ventromental plates very wide (about 1.5× width of mentum), almost straight and touching medially; mandible with 4 flattened inner teeth...*Axarus*

24' Ventromental plates curved, variable in lengths, close medially or broadly separated, but never touching medially 25

25(24) Ventromental plates distinctive: wide and strongly curved medially, separated at most by width of median mental tooth; seta subdentalis large with toothed inner margin, triangulum occipitale well developed .... *Goeldichironomus* in part

25' Ventromental plates normally curved, separated broadly medially, seta subdentalis single, triangulum occipitale not developed .... *Chironomus*

26(20) Line connecting ventromental plate and outer median mental teeth; mentum with all median teeth of the same color..........27

26' No line connecting ventromental plate and outer median mental teeth, which is often lighter in color than lateral teeth...........28

27(26) Ventromental plate elongate with anterior and posterior margins rather parallel, lateral apex rounded into a lobe, 1st lateral mental tooth subequal to 2nd lateral.............. *Endochironomus*

27' Ventromental plate not parallel sided, lateral apex not rounded into a lobe, truncate, 1st lateral mental tooth shorter than 2nd lateral tooth... *Endotribelos* in part

28(26) Mentum with median teeth of mentum light, separated by narrow notches, 1st lateral tooth not shorter than adjacent teeth, first 3 pairs of lateral teeth darker and larger than outer lateral ones that are fine and pale; V-shaped gap between 2nd and 3rd lateral teeth, ventromental plates coarse striated, anterior margin at most weakly crenulated..................................... *Pelomus*

28' Mentum with median teeth of mentum well separated, light, 2nd lateral tooth very long, 1st lateral tooth shorter than outer median and 2nd lateral tooth and may be fused with 2nd lateral, ventromental plates with very coarse striae reaching anterior margin... *Paratendipes*

*Modified after Epler (2001), Brooks et al. (2007), Epler et al. (2013), and Silva et al. (2018).

It is important to note that, in this book, we are dealing with parataxonomy; therefore, the identification of characters may not follow real morphological characters. *Apedilum*, for example, has 6 lateral teeth with the 1st one being very small and appressed to the 2nd lateral tooth. In subfossil material, however, the 1st lateral tooth is often worn down and is not visible or appears as a notch on the 1st lateral tooth, which is in fact the 2nd one; consequently, in the key, we may state that the specimen has 5 lateral teeth.

APEDILUM TOWNES

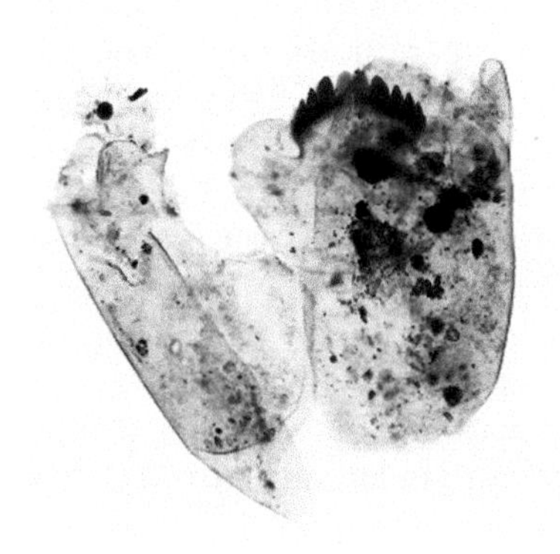

Diagnosis. Mentum with pale bifid median tooth and 6 lateral teeth (Figure 4.2a,b); 1st lateral tooth shorter and appressed to median teeth (hardly visible in worn down menta). Ventromental plates separated medially by width of paired median mental tooth, curved and finely striated. Premandible with 2 apical teeth and one small inner tooth. Mandible with a slender apical tooth, 3 inner teeth, one dorsal, and one surficial tooth. Antenna 6-segmented; Lauterborn organs alternate on 2nd and 3rd segments.

Remarks. The number of the mental teeth (and presence of frontoclypeal apotome) can separate *Apedilum* from the closely related *Microtendipes.*

Ecology and Distribution. Apedilum larvae are associated with submerged vegetation in ponds, littoral of lakes, and slowly moving rivers; recorded also from brackish water (Epler et al. 2013). *Apedilum* remains were common in lakes in Guatemala.

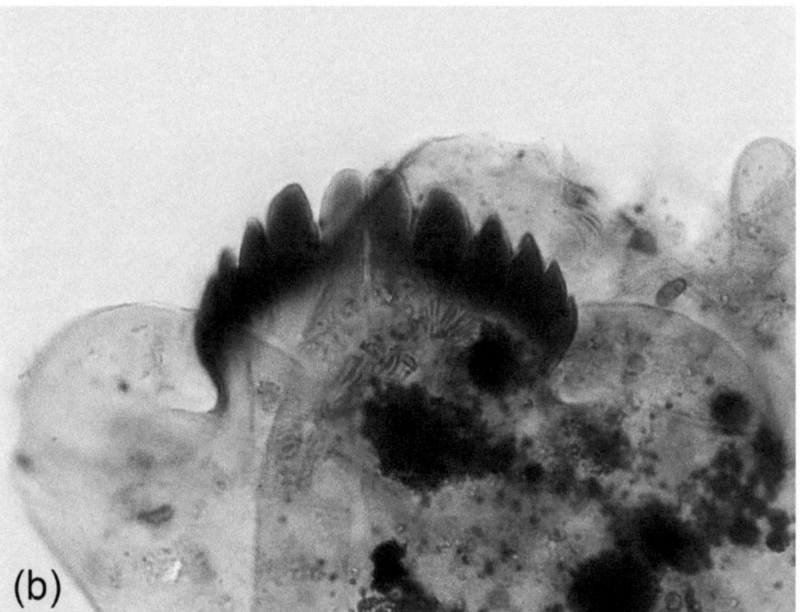

Figure 4.2 *Apedilum.* a—detail of head with labrum, mentum and mandible (note visible 1st lateral teeth and deformed right mandible), b—mentum (1st lateral teeth worn down, not visible).

AXARUS ROBACK

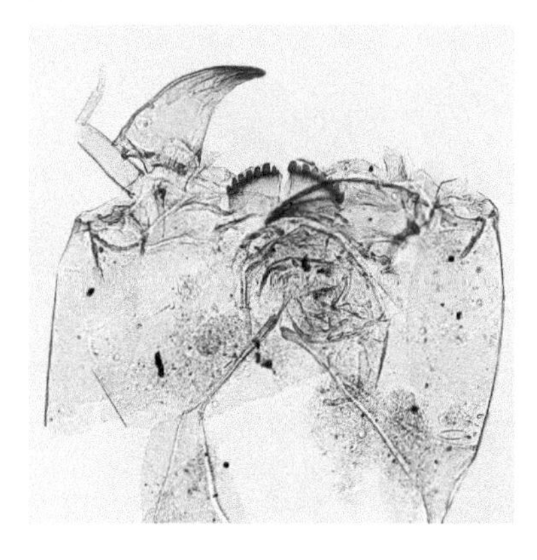

Diagnosis. Mentum with trifid median tooth, outer teeth small, 6 pairs of lateral teeth slightly decreasing in size (Figure 4.3a,d). Ventromental plates connecting medially, wide, nearly straight, with coarse striae restricted to a narrow transverse band. Premandible with long, slender apical tooth and 4 shorter pointy inner teeth. Mandible with slender apical tooth and 4 flattened inner teeth, in worn down specimens hardly visible, dorsal tooth absent (Figure 4.3c). Antenna 5-segmented; Lauterborn organs small, opposite on 2nd segment (Figure 4.3b).

Remarks. The combination of long, slender, medially almost connecting ventromental plates and 4 flattened inner mandibular teeth is distinctive for the genus.

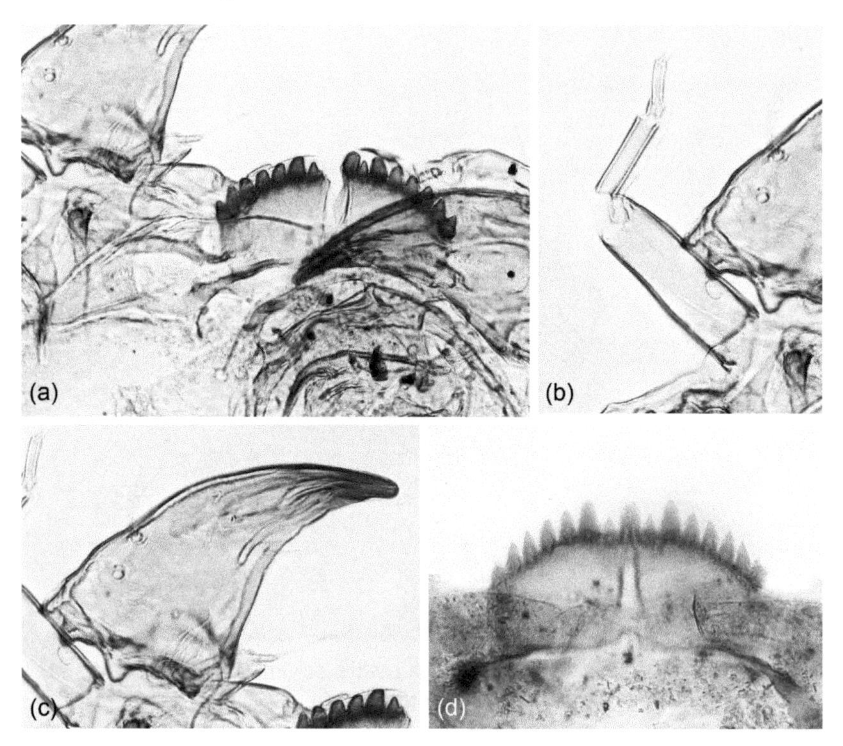

Figure 4.3 *Axarus.* a—mentum and ventromental plates, b—antenna, c—mandible, d—mentum.

Ecology and Distribution. Larvae of *Axarus* occur in soft sediments of littoral and sublittoral zone of lakes, also in slow flowing parts of rivers with varved clay substrates and decaying wood (Werle et al. 2004). Remains of the genus were rare in the Central American subfossil material.

BEARDIUS REISS & SUBLETTE

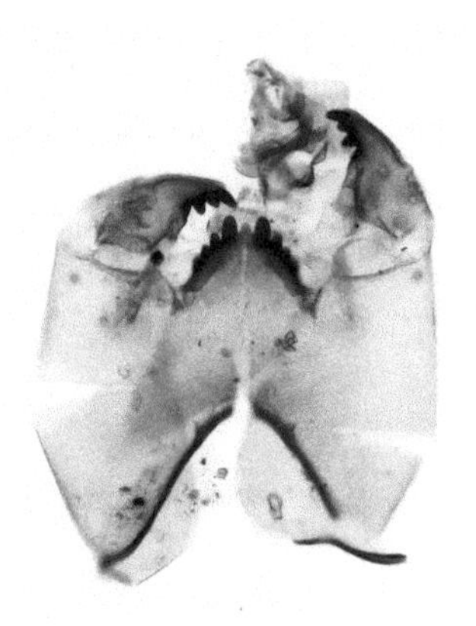

Diagnosis. Mentum with recessed simple pale median tooth; 5 pairs of lateral teeth declining in size laterally, 1st lateral tooth much larger than median tooth; ventromental plates short, curved, and coarsely striated, medially separated by distance between the first two lateral teeth (Figure 4.4a,c). Mandible with short apical tooth and 2 inner teeth, dorsal tooth present (Figure 4.4b,d). Premandible bifid. Antenna 7-segmented; Lauterborn organs large, alternate on 2nd and 3rd segments.

Remarks. In general, larval *Beardius* closely resemble *Oukuriella*, and can be distinguished from it by 2 inner teeth on mandible (3 in *Oukuriella*). Median mental tooth is always pale in *Beardius*, but can be both pale and dark in *Oukuriella*. Because the antenna and mandible are frequently missing in subfossil material, it may be problematic to

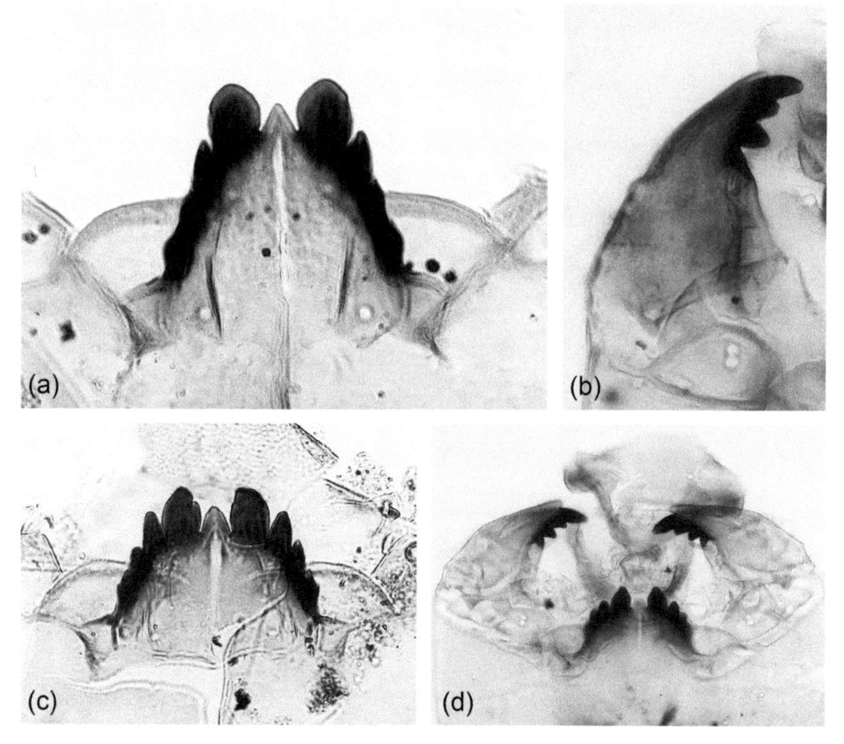

Figure 4.4 *Beardius*. *Beardius* type Chanmico: a—mentum, b—mandible; *Beardius* type Ipala: c—mentum, d—mentum and mandibles.

distinguish these two genera from each other. However, the shape of dorsoventral plates seems distinctive for distinguishing subfossil *Oukuriella* from *Beardius*. Lateral ends of the ventromental plates of *Oukuriella* point forward (anteriorly), while *Beardius* points either laterally or backward (L. Pinho, pers. comm.; see also figures in Trivinho-Strixino 2014).

KEY TO MORPHOTYPES

Due to the similarity of *Beardius* and *Oukuriella*, we provide a key including both genera.

1 Mentum with small dark recessed median tooth, much smaller than 1st lateral, mandible with 3 dark inner teeth.. *Oukuriella pinhoi*-type

1' Mentum with median tooth pale, about half the length of 1st lateral, lateral teeth dark, mandible with 2 inner teeth..............................2

2 Mentum with small, pale, distinctly triangular median tooth and seemingly 4 lateral teeth, 1st lateral dark, broad and distinctly round in shape, mentum strongly arched.............. *Beardius* type Chanmico

2' Mentum with pale median tooth and 5 dark lateral teeth, mentum not strongly arched...*Beardius* type Ipala

ADDITIONAL REMARKS TO MORPHOTYPES

Beardius type Chanmico (Figure 4.4a,b). Mentum has small, pale, distinctly triangular median tooth and 4 lateral teeth (most likely 5 but outermost 2 teeth fused), 1st lateral dark, broad, and round, mentum arched. Mandible with 2 inner teeth and a pale dorsal tooth.

Beardius type Ipala (Figure 4.4c,d). Mentum with pale median tooth, 5 dark lateral teeth (Figure 4.4c). Mandible with 2 dark inner teeth and a large dorsal tooth, variable in size.

There is a difference between the two morphotypes in the shape of ventromental plates: lateral ends of ventromental plates of *Beardius* type Ipala are tapering, pointed, while in *Beardius* type Chanmico lateral ends are oblique.

Ecology and Distribution. Essentially, a Neotropical genus. Larvae live in intermittent wetlands, marshes, ponds, streams, and rivers; some Neotropical species are associated with immersed macrophytes (Epler et al. 2013). *Beardius* type Chanmico is rare, while *Beardius* type Ipala is more common and was recorded throughout Central America.

CHIRONOMUS MEIGEN

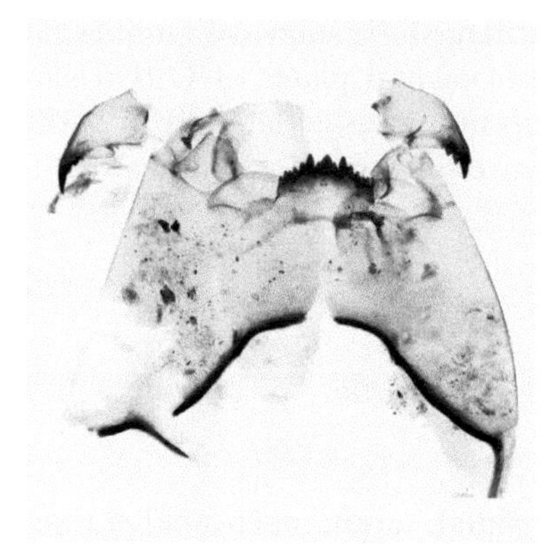

Diagnosis. Mentum with a characteristic pattern (Figure 4.5a,c); trifid median tooth with outer teeth usually much smaller than median tooth (half of it or even less); 6 pairs of lateral teeth, regularly decreasing in size, 1st lateral tooth long (usually the same size as median tooth) and often partly fused with the shorter 2nd lateral; the size of the 4th lateral tooth is important to distinguish morphotypes. Ventromental plates about as wide as mentum, with conspicuous striation at the base (Figure 4.5a). Premandible with 2 apical teeth, however, some Neotropical species may have 5 teeth. Mandible with basal row of radially arranged furrows (also found in *Benthalia*); usually with dark apical tooth, 2–3 inner teeth and a dorsal tooth (Figure 4.5a,d). Antenna 5-segmented; Lauterborn organs opposite on apex of 2nd segment (Figure 4.5b).

Remarks. Chironomus is closely related to *Dicrotendipes*, *Einfeldia*, *Goeldichironomus*, *Benthalia*, and *Kiefferulus*. In addition to the distinctive pattern of the three median teeth, *Chironomus* can be separated from the latter by the shape of the frons and ventromental plates.

KEY TO MORPHOTYPES

1 Mentum with lateral teeth decreasing gradually, 1st and 2nd lateral teeth are not fused, mandible with 3 inner teeth and 1 dorsal tooth ....*Chironomus* type Salto Grande

1' Mentum with 2nd lateral tooth shorter than 1st, closely associated with it or partially fused..2

2 Mentum with 4th lateral tooth smaller than 3rd and 5th lateral ....*Chironomus anthracinus*-type

2' Mentum with lateral teeth gradually decreasing, 4th lateral tooth longer than 5th..*Chironomus plumosus*-type

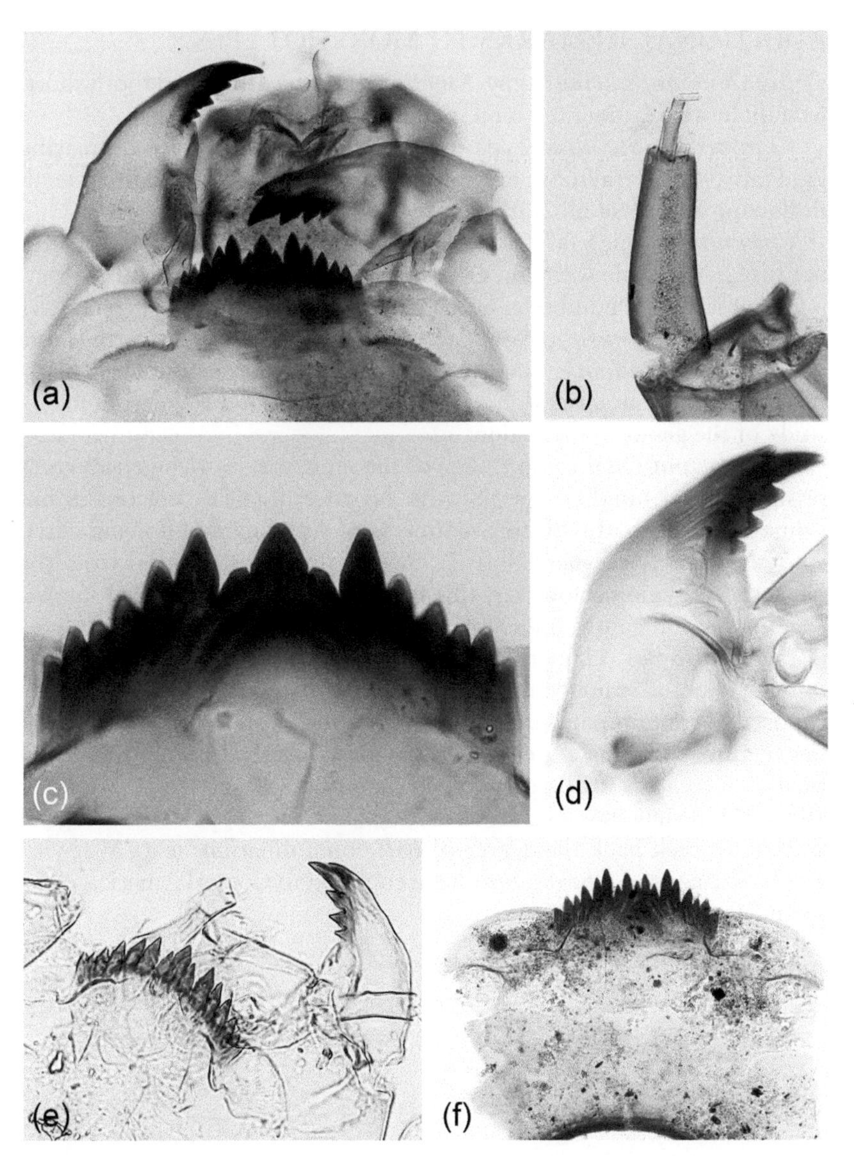

Figure 4.5 *Chironomus. Chironomus plumosus*-type: a—detail of head with mentum and mandible, b—antenna; *Chironomus anthracinus*-type: c—mentum, d—mandible; *Chironomus* type Salto Grande: e—mentum and mandible; f—buccal deformity (note 5 median teeth).

ADDITIONAL REMARKS TO MORPHOTYPES

Chironomus anthracinus-type. Mentum with the 4th lateral tooth short. Mandible with 2 inner teeth on mandible (Figure 4.5c–d).

Chironomus plumosus-type. Mentum with lateral teeth decreasing gradually, 4th lateral tooth longer than 5th. Mandible with 3 inner teeth (following Brooks et al. 2007; Figure 4.5a).

Chironomus type Salto Grande. Mentum with 1st and 2nd lateral teeth not fused. Mandible with 3 inner teeth and 1 dorsal tooth (Figure 4.5e).

There is a great number of head capsules with a mixture of characters of the *C. anthracinus*-type and *C. plumosus*-type. Considering the large number of *Chironomus* species worldwide, these remains most likely represent different species/types and their description requires complex study of the genus in the region.

Ecology and Distribution. One of the most species-rich genera comprising several hundred species with broad ecological preferences but primarily associated with soft sediments of standing and flowing nutrient-rich waters. Spies and Reiss (1996) catalogued 19 species from the Neotropical region. However, this number represents outdated knowledge and significantly underestimates actual diversity of *Chironomus* in the Neotropics (Hamerlik et al. 2018b and references therein). Due to their tolerance and feeding habit, *Chironomus* larvae can be directly exposed to contaminants in sediments and thus prone to buccal deformities (Figure 4.5f; Janssens et al. 1992). Consequently, they are frequently used as indicators of toxic contaminants in freshwaters (e.g., Warwick 1985; for review, see Nazarova 2000) and also in paleolimnological studies to track back the degree of past contaminations (e.g., Warwick 1980). Remains of *Chironomus* were common in Central American lake sediments.

CLADOPELMA KIEFFER

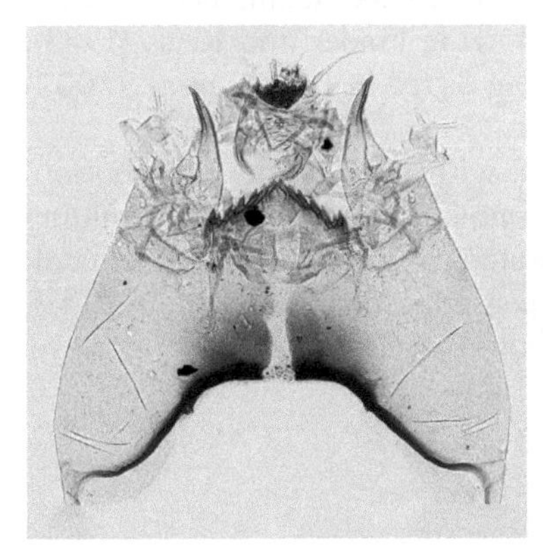

Diagnosis. Mentum strongly arched, double median tooth (may seem medially notched single median tooth), 7 pairs of lateral teeth (Figure 4.6a–b). Three outermost teeth offset and clustered, with a characteristic pattern: 5th and 7th lateral teeth are short, while the 6th is higher and broader. Ventromental plates broad, with fine striation, anterior margin often crenulated. Premandible with 2 teeth. Mandible with apical tooth and 1–2 flat inner teeth, dorsal tooth absent (Figure 4.6a). Antenna 5-segmented; Lauterborn organs indistinguishable.

Remarks. Mentum with enlarged outer teeth; with a notched or bifid median tooth, not projecting strongly forward from the mentum will distinguish *Cladopelma* from other Chironomini including the most similar genera, *Microchironomus* and *Cryptotendipes*.

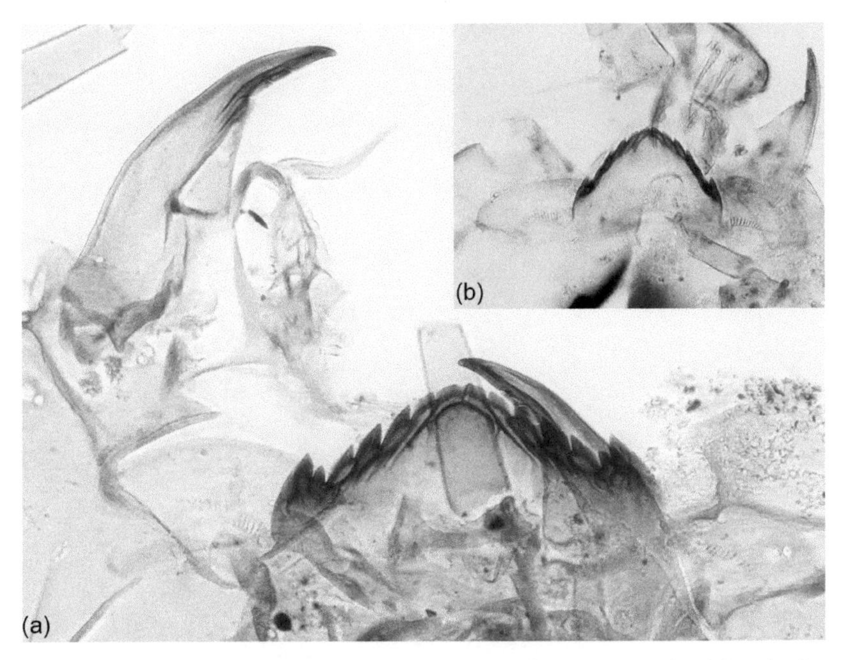

Figure 4.6 *Cladopelma*. a—mentum and mandible, b—detail of head of a young instar.

The morphotype found in Central American lake sediments was identified as *Cladopelma lateralis*-type following Pinder and Reiss (1983). The type is slightly different form the one listed in Brooks et al. (2007) by the broader median teeth.

Ecology and Distribution. Genus with worldwide distribution. Larvae are eurytopic and live in various aquatic environments including ponds, lakes, streams, rivers, brackish waters, and hot springs (Epler et al. 2013). Some species tolerate low oxygen concentrations (Epler 2001). Common in surface sediments throughout Central America.

CRYPTOCHIRONOMUS KIEFFER

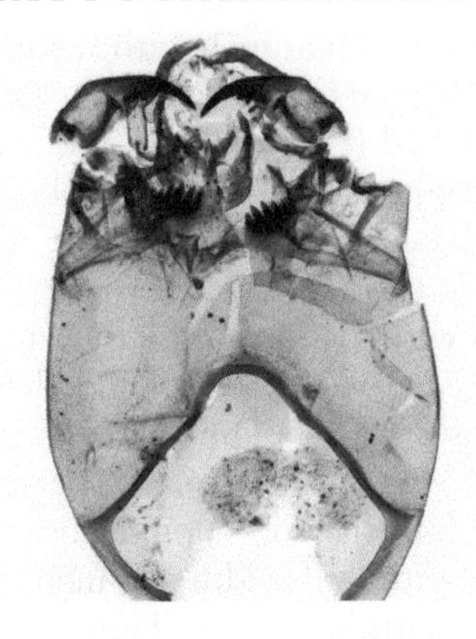

Diagnosis. Mentum concave with pale, broadly domed median tooth and 6–7 pairs of dark lateral teeth; 1st lateral merged with median tooth; outermost lateral tooth notched apically (Figure 4.7a,b). Ventromental plates very large, wider than mentum, with fine striation and smooth anterior margin (Figure 4.7b). Premandible with 4–6 teeth declining in size from apex to base. Mandible with long slender apical tooth and 2 pointed inner teeth, dorsal tooth absent. Antenna 5-segmented; Lauterborn organs absent.

Remarks. The 5-segmented antenna and the shape of the mentum, with a broad, pale median tooth, 6 or 7 pairs of obliquely arranged, dark lateral teeth distinguish *Cryptochironomus* larvae from all other Chironomini.

Ecology and Distribution. Genus with very broad ecological requirements; larvae live in lake littoral, streams, and rivers, where they prefer sandy substrate. Relatively rare in Central American lakes.

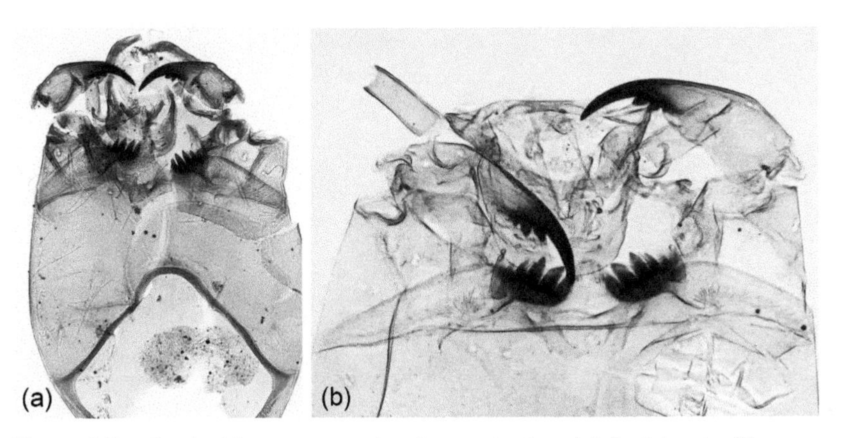

Figure 4.7 *Cryptochironomus.* a—head capsule, b—detail of head with mentum, ventromental plates and mandible with long apical teeth.

DICROTENDIPES KIEFFER

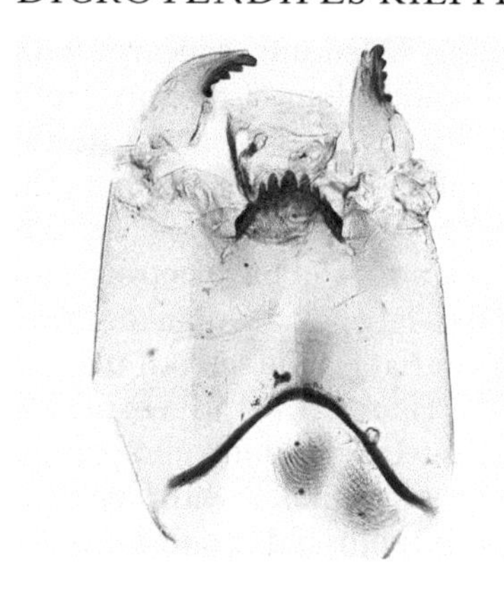

Diagnosis. Mentum with simple, robust median tooth that can be laterally notched (Figure 4.8d), about the same height as 1st lateral; 5–6 lateral teeth, 1st lateral is the highest, 2nd lateral is shorter and often partially merged with the 1st lateral, outermost lateral can be partly to completely fused to form a bulging lobe (Figure 4.8f,g). Ventromental plates narrow, visibly fan-shaped, completely striated, about half-width of mentum; anterior margin smooth (Figure 4.8b) to coarsely crenulated (Figure 4.8e). Premandible with bifid apical tooth and blunt basal tooth. Mandible with prominent apical tooth, pale dorsal tooth, 3 inner teeth; 1–2 small surficial teeth may be present, outer margin frequently crenulated (Figure 4.8h–i). Antenna 5-segmented; Lauterborn organs opposite on apex of 2nd segment.

Remarks. Larvae of *Dicrotendipes* are somewhat similar to those of *Goeldichironomus*; however, the placement and shape of the ventromental plates separate these two genera. A useful character to distinguish the genus is that the frons is separated from clypeus, and there is an oval to round or linear frontal pit, antero-median mark on the frons, anterior margin of frons may be crenulated/tuberculated.

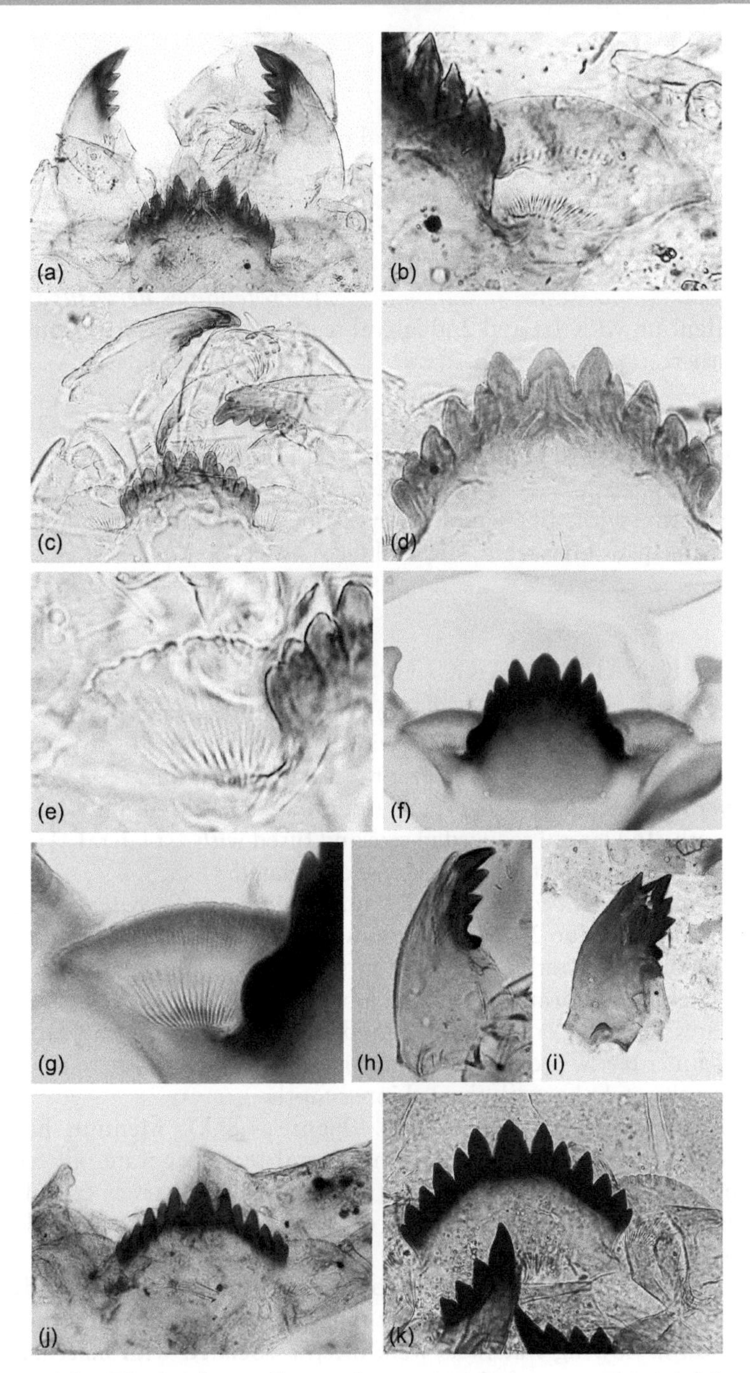

Figure 4.8 *Dicrotendipes. Dicrotendipes notatus*-type: a—mentum and mandible, b—ventromental plate; *Dicrotendipes nervosus*-type A: c—mentum and mandible, d—detail of mentum, e—ventromental plate; *Dicrotendipes nervosus*-type B: f—mentum, g—ventromental plates, h—normal mandible, i—worn down mandible; *Dicrotendipes nervosus*-type C: j—mentum, k—mentum and mandible.

KEY TO MORPHOTYPES

1 Anterior margin of ventromental plates smooth ....*Dicrotendipes notatus*-type
1' Anterior margin of ventromental plates crenulated 2

2 Mentum with 2nd lateral teeth appressed to 1st lateral ....*Dicrotendipes nervosus*-type A
2' Mentum with 1st and 2nd lateral teeth clearly separated from each other 3

3 Mentum with, 5th and 6th teeth are fused; mandible with 3 inner teeth, 3rd tooth reduced, surficial teeth present ....*Dicrotendipes nervosus*-type B
3' Mentum with, all 6 lateral teeth distinctly separated; mandible with 3 subequal inner teeth, surficial teeth absent ....*Dicrotendipes nervosus*-type C

ADDITIONAL REMARKS TO MORPHOTYPES

Dicrotendipes notatus-type (Figure 4.8a,b). Anterior margin of ventromental plates smooth. Mandible with 3 subequal inner teeth. Mentum with 6 lateral teeth, 2nd lateral tooth fused with 1st one.

Dicrotendipes nervosus-type A (Figure 4.8c–e). Mentum with 6 lateral teeth, 2nd lateral reduced in size and fused with 1st lateral. Anterior margin of ventromental plates coarse crenulated.

Dicrotendipes nervosus-type B (Figure 4.8f–i; see *Dicrotendipes* sp. 1 in Trivinho-Strixino 2014). Mentum with 1st and 2nd lateral teeth separate, teeth significantly decreasing in size laterally, outermost lateral is completely fused forming a bulging lobe. Anterior margin of ventromental plates with fine crenulation. Mandible with prominent apical tooth and 3 inner teeth, 3rd tooth minute; 1–2 surficial teeth; there is a dark lobe on the mola beneath the 3rd inner tooth.

Dicrotendipes nervosus-type C (Figure 4.8j,k). Mentum has all 6 lateral teeth distinct, 1st and 2nd lateral teeth separate, all mental teeth dark. Anterior margin of ventromental plates coarsely crenulated. Apical and inner teeth of mandible dark, 3 inner teeth of subequal size, surficial teeth absent.

Ecology and Distribution. Dicrotendipes is a genus with worldwide distribution. Larvae predominantly inhabit littoral of lakes but may be common in flowing waters as well; often associated with macrophytes (Epler 2001). In general, *Dicrotendipes* was common in lake sediments with *D. nervosus*-type B as the most frequently recorded.

ENDOCHIRONOMUS KIEFFER

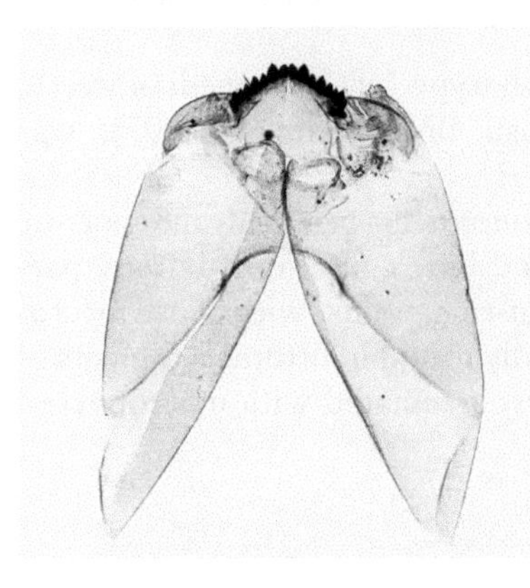

Diagnosis. Mentum with distinct dorsomentum with 3–4 median teeth, the central ones shorter than the outer pair; 6 pairs of lateral teeth are delineated from median teeth by a distinct line linking the inner margin of ventromental plates to the bases of the outer median tooth (Figure 4.9). Ventromental plates elongate, parallel sided, outer margin strongly curved and rounded into a lobe; striae interrupted medially. Premandible with 3 teeth. Mandible with slender apical tooth and 3–4 inner teeth; dorsal tooth present, not extending beyond the outer margin. Antenna 5-segmented diminishing in size distally; Lauterborn organs about as long as 3rd segment, opposite on apex of 2nd segment.

Remarks. Endochironomus is distinct by a mentum with 4 (occasionally 3) median teeth separated from the lateral part of the mentum by a clear line; ventromental plates with parallel anterior and posterior margins for most of their length, and lateral apex rounded into a lobe. The frons separated from clypeus and mandible without notch

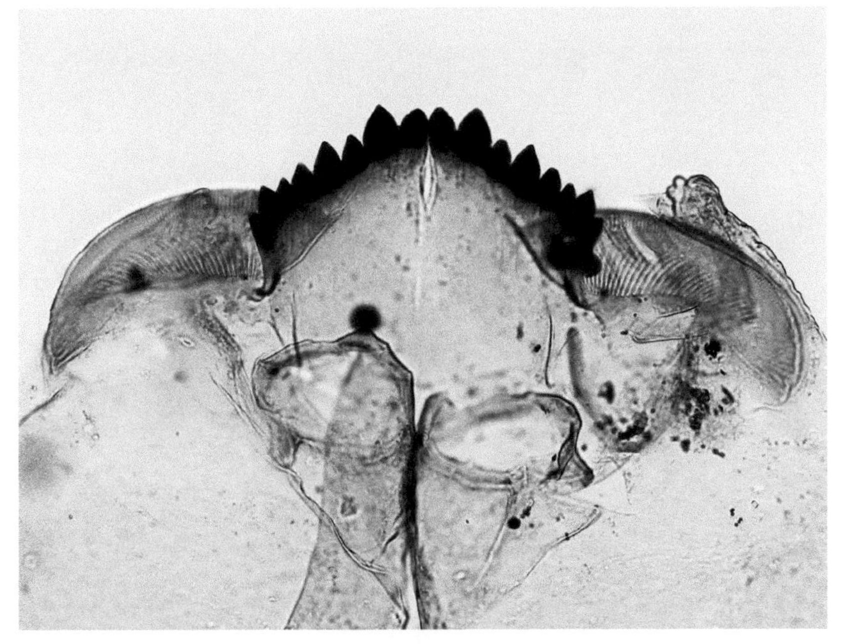

Figure 4.9 *Endochironomus albipennis*-type. Mentum and ventromental plates.

allow separation of *Endochironomus* from *Endotribelos*, *Imparipecten*, *Phaenopsectra*, and *Tribelos*.

In Central America, only a morphotype with 4 median teeth, *Endochironomus albipennis*-type (sensu Brooks et al. 2007), was identified.

Ecology and Distribution. Endochironomus is primarily a temperate Holarctic genus. Larvae live in lentic and lotic conditions, but are particularly characteristic for small, nutrient-rich ponds, with dense macrophyte stands. Subfossil remains are usually found in littoral sediments of mesotrophic and eutrophic lentic waters, associated with macrophytes. Rare in Central American waterbodies.

ENDOTRIBELOS GRODHAUS

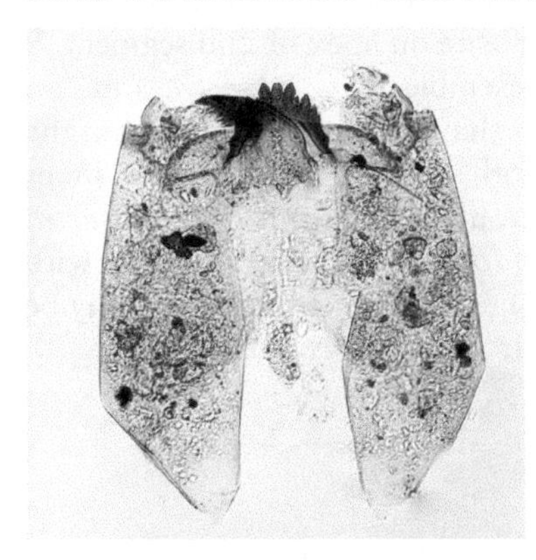

Diagnosis. Mentum with distinct dorsomentum bearing 3–4 median teeth (Figure 4.10a,c), central tooth of those with 3 teeth sometimes notched (Figure 4.10b); 6 pairs of lateral teeth are present with the 1st much lower than median or 2nd; 6th lateral may be reduced or absent. When 4 median teeth are present, the 2 central teeth are subequal or smaller than the outer median teeth (Figure 4.10c,d). Ventromental plates medially connected to bases of outer ventromental median teeth; relatively narrow and strongly curved; striae distinct and continuous. Premandible with 2 long apical teeth and 1 blunt basal tooth. Mandible with strong apical tooth and 3–4 inner teeth (Figure 4.10b,d); pale dorsal tooth present. In some species, there is a distinct gap between the mola (the inner margin of the mandibular

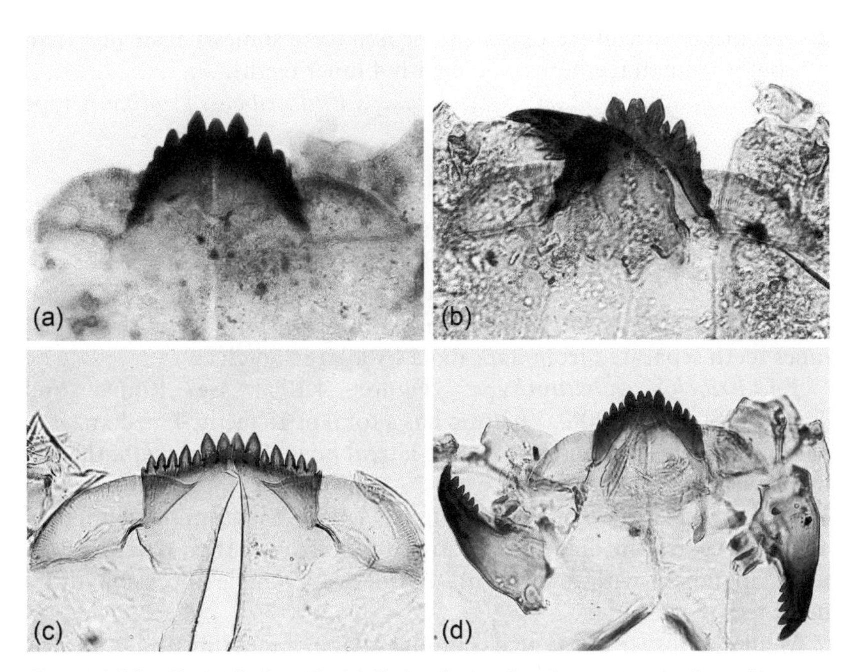

Figure 4.10 *Endotribelos. Endotribelos hesperium*-type: a—mentum, b—morphotype with notched median tooth (note the gap between mola and the 1st inner tooth mandible); *Endotribelos albatum*-type: c—mentum; *Endotribelos grodhausi*-type: d—detail of head with mentum and mandible.

base) and the inner mandibular tooth (Figure 4.10b). Antenna 5-segmented; Lauterborn organs minute, opposite on apex of 2nd segment.

Remarks. Larvae of *Endotribelos* resemble *Phaenopsectra* and can be distinguished by the arched mentum shape and the gap between the mola and the 1st inner mandibular tooth. The gap between 1st inner mandibular tooth and mola is, however, also present in *Phaenopsectra flavipes* and could be confused with *Endotribelos hesperium*-type with such a gap, which has similar mentum, but only 3 median teeth (may be notched) as opposed to 4 median teeth in *Phaenopsectra flavipes*.

KEY TO MORPHOTYPES

1 Mentum strongly arched with 3 subequal median teeth (central one may be notched) and 5 lateral teeth... .... *Endotribelos hesperium*-type

1' Mentum with 4 median teeth...2

2 Mentum with 16 teeth, 4 median teeth about the same size, higher than the lateral ones, 6 lateral teeth, the outermost one minute .... *Endotribelos albatum*-type

2' Mentum with 14 teeth, central median teeth somewhat smaller than outer median teeth, mandible with 4 inner teeth.............................. .... *Endotribelos grodhausi*-type

ADDITIONAL REMARKS TO MORPHOTYPES

Endotribelos hesperium-type (Figure 4.10a,b). Mentum is strongly arched, with odd number of teeth, 3 median teeth of subequal size, central tooth notched, higher than lateral ones, 5 lateral teeth, 1st lateral is small and pressed to the outer medial tooth. Mandible with 3 equal inner teeth separated from dark mola by a large notch.

Endotribelos albatum-type (Figure 4.10c; see Roque and Trivinho-Strixino 2008). Mentum has a total of 16 teeth, 4 median teeth about the same size, higher than the lateral ones, 6 lateral teeth, the last one lower than the rest.

Endotribelos grodhausi-type (Figure 4.10d). Mentum has a total of 14 teeth, central medial teeth shorter and narrower than outer median teeth, mandible with 4 (in Central American samples occasionally 5) inner teeth.

While *E. hesperium*-type has unique shape of mentum and is hard to confuse with other taxa, the two other types with 4 median mental teeth (*grodhausi*- and *albatum*-types) superficially resemble *Endochironomus* and can be distinguished from it by the shape of ventromental plates that are strongly curved and rounded into a lobe in *Endochironomus*.

Ecology and Distribution. Larvae of *Endotribelos* are generally associated with macrophytes and can be important colonizers of immersed leaf litter. Some species are widespread and common dwellers of fruit tissues of several tree species in forested streams (Roque et al. 2005). *Endotribelos* remains were rare in Central America lake-surface sediments.

FISSIMENTUM CRANSTON & NOLTE

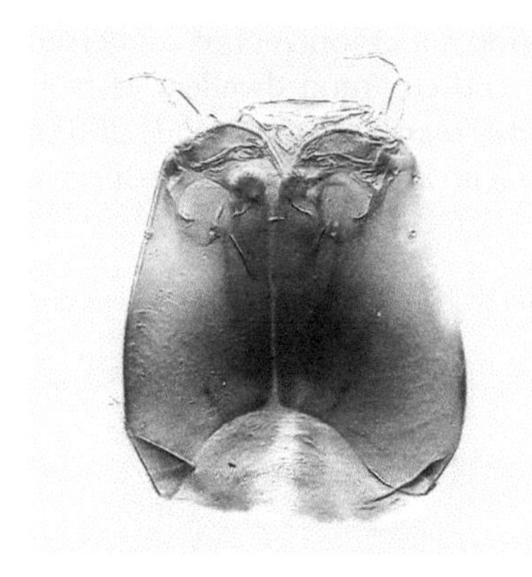

Diagnosis. Mentum with deeply sunken pair of median teeth is unique and easily distinguish *Fissimentum* from other genera. Ventromental plates elongate with smooth margin (Figure 4.11a). Premandible bifid. Mandible with robust apical tooth and 3 small uniform inner teeth; dorsal tooth absent; seta subdentalis large, sinuous, extended out to apical tooth (Figure 4.11b). Antenna 6-segmented; Lauterborn organs small to moderate, alternate on 2nd and 3rd segments.

Remarks. The mentum with deeply recessed median teeth is distinctive for *Fissimentum* and *Hyporhygma*. The latter can be separated from *Fissimentum* by the 5-segmented antenna. One morphotype, *Fissimentum desiccatum*-type (see Cranston and Nolte 1996), was identified.

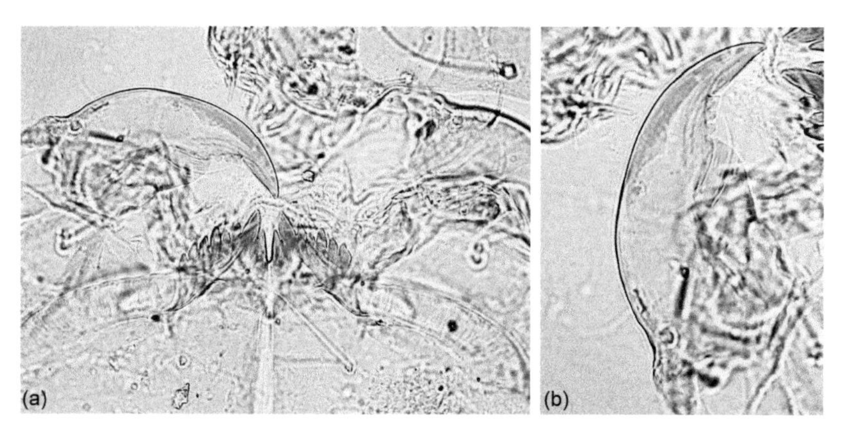

Figure 4.11 *Fissimentum desiccatum*-type. a—mentum and mandible, b—detail of mandible with seta subdentalis.

Ecology and Distribution. Larvae of *Fissimentum* live in soft and sandy substrata of rivers and streams. *Fissimentum desiccatum*, as the name indicates, can be resistant to some extent to drought (Cranston and Nolte 1996). Owing to the rheophilic nature of larval *Fissimentum*, head capsules are rare in the subfossil material.

GOELDICHIRONOMUS FITTKAU

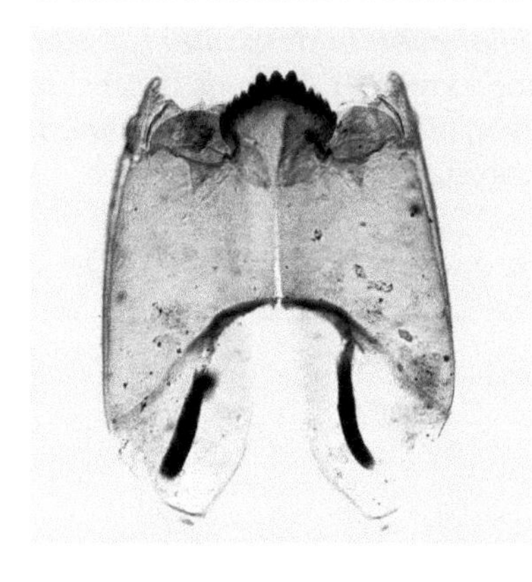

Diagnosis. Median mental tooth trifid or single (in that case, laterally crenate); number of lateral teeth may vary but usually with 6–7 pairs (Figure 4.12c,d); 2nd lateral may be reduced and fused with 1st lateral; 4th lateral subequal or significantly lower than 3rd and 5th. Ventromental plates characteristic, wide, ventrally tilted, pointing downward and almost touching medially (Figure 4.12a). Ventral occipital margin heavily sclerotized, with a distinct, wide triangulum occipitale (Figure 4.12f). Premandible has 2–3 teeth. Mandible with apical tooth followed by 3 inner teeth; pale dorsal tooth present; seta subdentalis remarkably large, comb-like (Figure 4.12h). Antenna 5-segmented, segments are diminishing in size distally; Lauterborn organs opposite on apex of 2nd segment (Figure 4.12g).

Remarks. The combination of the distinctive shape of ventromental plates, well-developed triangulum occipitale, and the unique, large, sickle-shaped seta subdentalis, which is toothed along its inner margin, will separate *Goeldichironomus* from other Chironomini.

KEY TO MORPHOTYPES

1 Mentum with median tooth small, deeply sunken, 1st and 2nd lateral teeth partly fused.................. *Goeldichironomus* type Olomega

1' Mentum with median teeth longer or subequal to 1st lateral one.......2

2 Mentum with 15 teeth, 4th lateral tooth minute .... *Goeldichironomus carus*-type

2' Mentum with 13 teeth, 4th lateral tooth subequal to adjecent teeth 3

3 Mentum with median tooth high and narrow, subequal in size to 1st lateral; 2nd lateral tooth adpressed toward 1st lateral; with a broad gap between 2nd and 3rd lateral teeth... .... *Goeldichironomus holoprasinus*-type

3' Mentum with median tooth narrower than 1st lateral, lateral teeth subequal in size, outermost lateral tooth large and pointing outward .... *Goeldichironomus amazonicus*-type

ADDITIONAL REMARKS TO MORPHOTYPES

Goeldichironomus type Olomega (Figure 4.12a,b). Mentum with recessed small median tooth, 2nd lateral fused with the large 1st lateral, 4–6 lateral teeth reduced and merged. Mandible with 4 inner teeth and 1 dorsal tooth, all dark, outer margin with extensive tuberculation, mola extended to a large rounded lobe.

Goeldichironomus carus-type (Figure 4.12c). Mentum with 15 teeth in total, medium tooth broader than 1st lateral, notched laterally, 4th lateral minute, smaller than 3rd and 5th tooth, outermost lateral tooth minute. Larvae of this type with 4th lateral minute are identified as *G. pictus* group in Epler et al. (2013). However, *Goeldichironomus pictus* has more lateral teeth. Named *carus*-type because of the similarity to *Goeldichironomus carus*, previously sampled in Central America (see Epler 2001).

Goeldichironomus holoprasinus-type (Figure 4.12d). Mentum with a total of 13 teeth, median tooth high and narrow with a notch on each side, 1st lateral of the same heights and width as median tooth; 2nd lateral teeth adpressed toward 1st one; with a broad gap between the 2nd and 3rd lateral teeth; outer 3rd to 6th lateral teeth pointing somehow outward. Larvae of *G. holoprasinus* have an additional dorsal tooth near the inner teeth of the mandible (see Epler 2001); this feature was not visible in the Central American material.

Goeldichironomus amazonicus-type (Figure 4.12e–h). Mentum with 13 teeth altogether, median tooth narrower than 1st lateral, lateral teeth subequal, outermost lateral tooth large and pointing outward.

Ecology and Distribution. Goeldichironomus is a speciose and primarily a tropical/subtropical genus, with most of the species occurring in Central and South America. Larvae are found mostly in soft sediments of standing waters, but can occur in or on plants and in floating algal mats of vegetation and wood. Larvae may occur in waters ranging from oligotrophy to hypereutrophy, from fresh to brackish water (Epler 2001). *Goeldichironomus* remains were frequent and abundant in Central American surface sediments. *Goeldichironomus amazonicus*-type and *Goeldichironomus carus*-type were the most common ones.

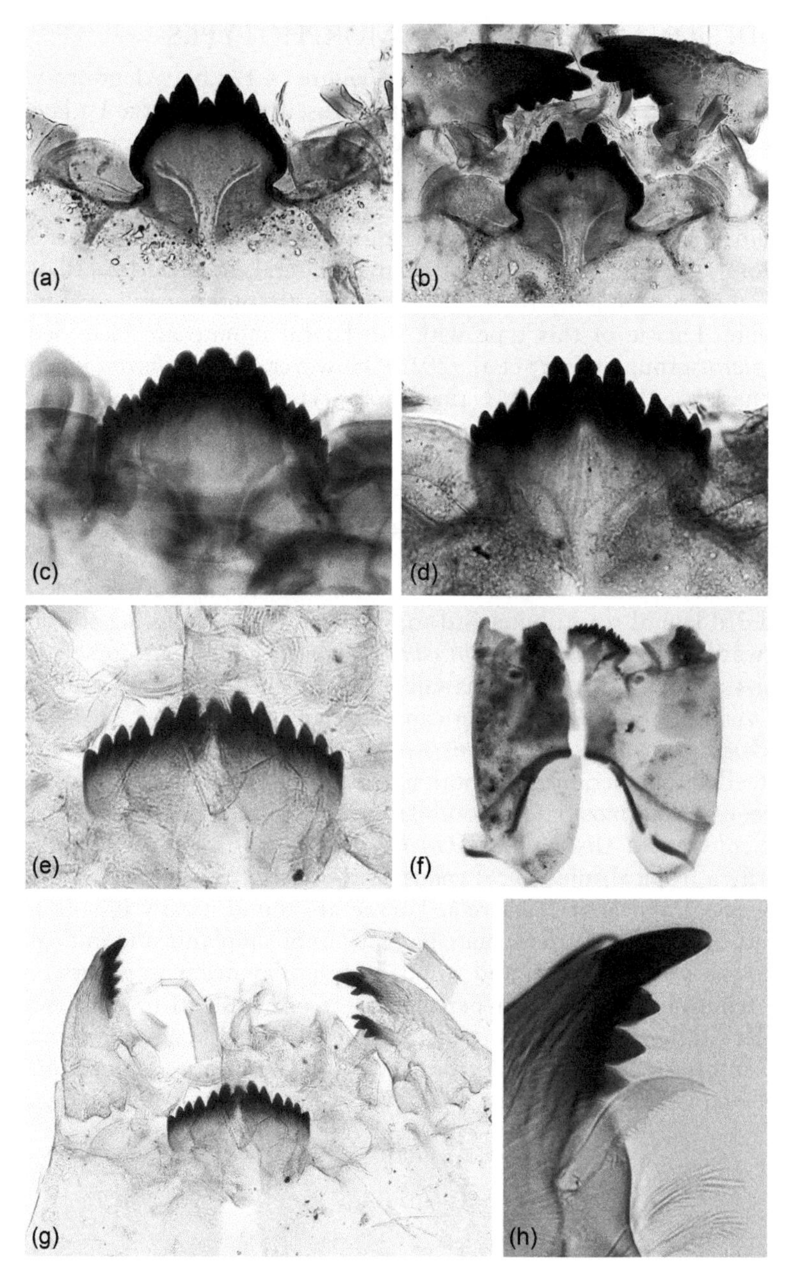

Figure 4.12 *Goeldichironomus*. *Goeldichironomus Olomega*-type: a—mentum (normal), b—mentum and mandible (worn down); *Goeldichironomus carus*-type: c—mentum; *Goeldichironomus holoprasinus*-type: d—mentum; *Goeldichironomus amazonicus*-type: e—mentum, f—head capsule showing well-developed triangulum occipitale, g—detail of head with mentum, mandible, premandible, and antenna, h—detail of mandible showing toothed seta subdentalis.

HARNISCHIA COMPLEX TYPE A

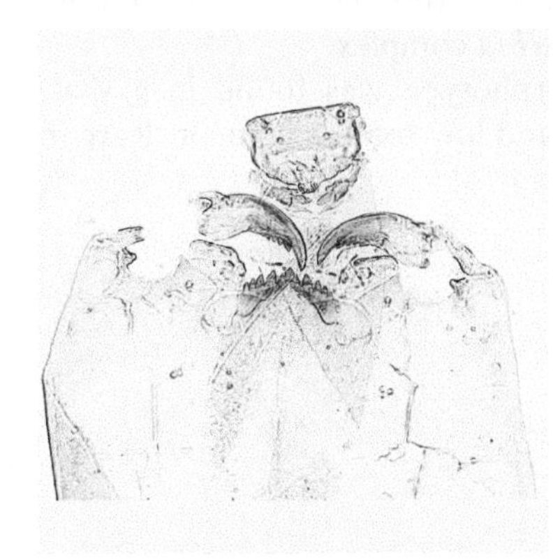

Diagnosis. Mentum has a trifid median tooth of triangular shape, outer teeth about half the size of the central tooth and partly fused with it; 5 pairs of lateral teeth, regularly decreasing in size, outermost minute. Ventromental plates shorter than mentum, curved, broadly separated medially. Premandible with 2 apical teeth and 1 basal tooth. Mandible with very long, curved apical tooth, 3 inner teeth diminishing in size toward the base, dorsal tooth absents; seta submenti long, reaching until the half of the apical tooth; apical half of mandible brownish, the same color as the mentum. Antenna missing in the recorded specimens (Figure 4.13).

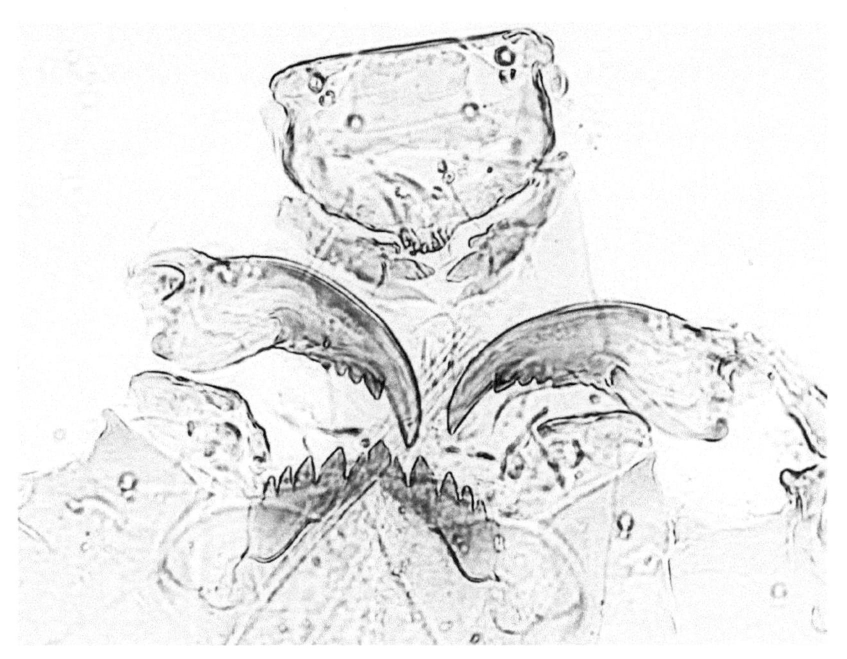

Figure 4.13 *Harnischia*-complex-type A. Detail of head showing mentum, mandible, ventromental plates, and premandible.

Remarks. The morphological features of this morphotype indicate that it could be a member of the *Harnischia* complex.

Ecology and Distribution. The morphotype was found in a shallow waterbody with high conductivity and low oxygen content. Rare in Central American lakes.

LAUTERBORNIELLA THIENEMANN & BAUSE

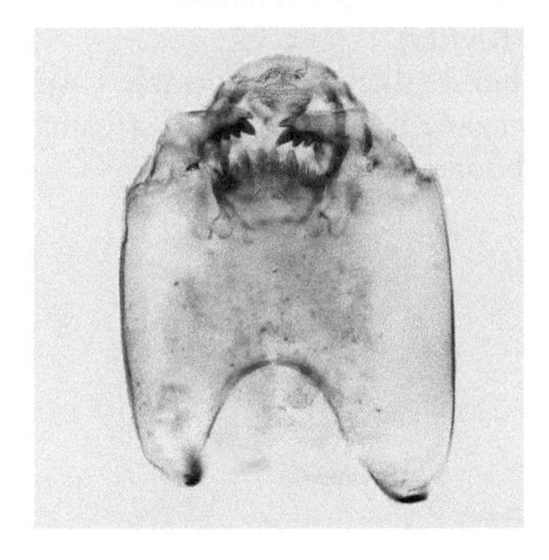

Diagnosis. Mentum uniformly brown, weakly pigmented, with a pair of rounded median teeth; 6 pairs of lateral teeth, 1st small, 2nd about as high as median teeth, outer lateral teeth gradually decreasing in size (Figure 4.14a). Ventromental plates wide, nearly triangular, with straight anterior margin, almost meeting medially, basal $^{2}/_{3}$ striated, anterior $^{1}/_{3}$ without striation; setae submenti large, plumose (Figure 4.14c). Premandible with 3 teeth and 1 blunt basal tooth. Mandible with uniformly brown apical tooth and 2 inner teeth, very strong pale dorsal tooth present (Figure 4.14a). Antenna 6-segmented; Lauterborn organs large, alternate on apex of 2nd, and on short pedestal subapically on 3rd segment.

Remarks. Lauterborniella can be distinguished from very similar *Zavreliella* by the arrangement of 1st and 2nd lateral teeth of the mentum, which are distinctly separate in *Lauterborniella*, while partially fused in *Zavreliella*. Setae submenti are also distinctive, branched in

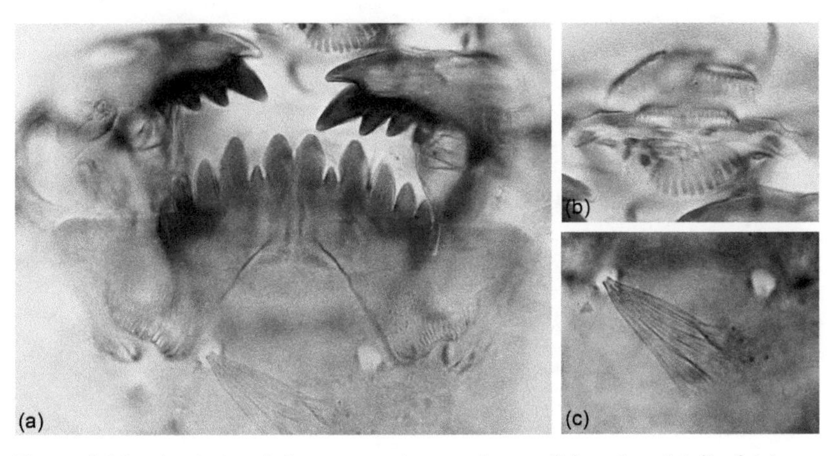

Figure 4.14 *Lauterborniella.* a—mentum and mandibles, b—detail of labrum showing palmate SI setae, labral lamellae, and pecten epipharyngis (consisting of 3 serrate lobes), c—branched seta submenti.

Lauterborniella and simple in *Zavreliella*. The genus is monotypic with the only known species *L. agrayloides* (Kieffer).

Ecology and Distribution. Larvae build silken transportable cases and live among vegetation in ponds and slow-moving areas of streams. Rather rare in Central American waterbodies.

MICROTENDIPES KIEFFER

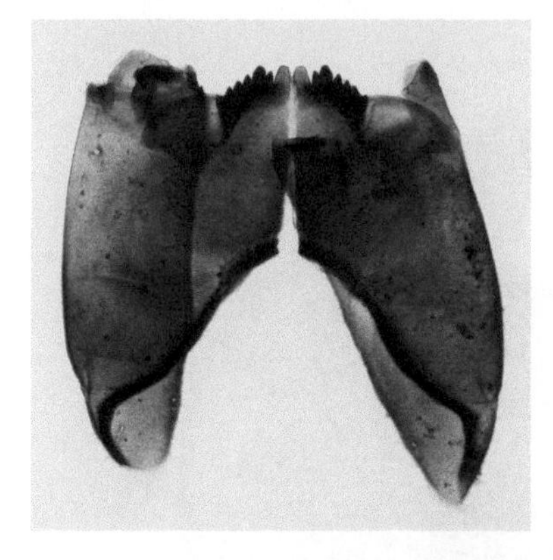

Diagnosis. Mentum with three median teeth (in *Microtendipes pedellus*-type, the central tooth is minute and depressed, thus appearing bifid; Figure 4.15a), usually paler than lateral teeth, occasionally as dark as 6 lateral teeth, 1st lateral tooth reduced, fused with 2nd lateral. Ventromental plates strongly curved with continuous coarse striation; a line connecting the upper inner margin of ventromental plates and the base of 2nd lateral tooth divides the mentum into central and lateral parts. Premandible with 3 (*M. pedellus*-type) or 5 teeth (*M. rydalensis*-type). Mandible with short apical tooth and 3 inner teeth, pale dorsal tooth present. Antenna 6-segmented; Lauterborn organs alternate on apices of 2nd and 3rd segments.

Remarks. Both *Paratendipes* and *Omisus* resemble *Microtendipes* and can be distinguished from it by the mandible with 4 median teeth (3 in *Microtendipes*). The number of mental teeth (and the frons separated from clypeus) will separate *Microtendipes* from the similar *Apedilum.*

KEY TO MORPHOTYPES

1 Median teeth appearing bifid (a vestigial tooth is present between them though), median teeth usually pale, occasionally dark .... *Microtendipes pedellus*-type

1' Three subequal median teeth, always paler than lateral teeth .... *Microtendipes rydalensis*-type

ADDITIONAL REMARKS TO MORPHOTYPES

Two morphotypes were distinguished following Epler et al. (2013).

Microtendipes pedellus-type (Figure 4.15a). Mentum with median teeth appearing bifid, but there is a hardly visible vestigial tooth between them. Median teeth usually pale, however, in some species, may be dark.

Microtendipes rydalensis-type (Figure 4.15b). Mentum with pale and trifid median tooth, teeth are subequal, occasionally the central one slightly smaller; lateral teeth darker than median ones.

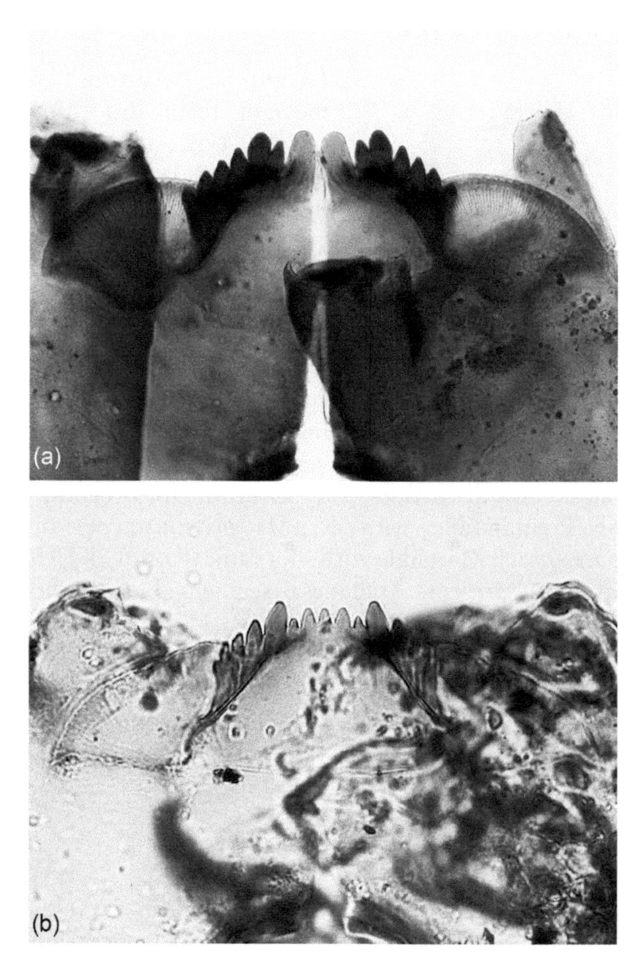

Figure 4.15 *Microtendipes. Microtendipes pedellus*-type: a—mentum; *Microtendipes rydalensis*-type: b—mentum.

Ecology and Distribution. Larvae of *Microtendipes* are characteristic of littoral and sublittoral sediments of lakes and ponds but can also occur in streams and rivers, where they are associated with soft sediment and macrophytes. In Central America, only found in high elevation lakes.

NILOTHAUMA KIEFFER

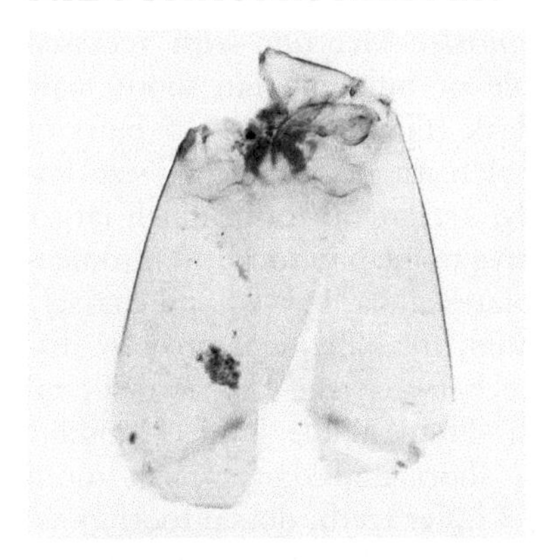

Diagnosis. Mentum pale, rather horizontal, all teeth uniform in color, median tooth bifid, subequal or slightly shorter than 1st lateral, with 6 lateral teeth evenly decreasing in size laterally (Figure 4.16b). Ventromental plate slightly wider than mentum with only basal half striated. Premandible with 3 teeth. Mandible of characteristic shape, with pale, long, and slender apical tooth, up to 4 inner teeth on different focal plane (Figure 4.16b), moderate dorsal tooth; all teeth pale or inner teeth slightly darker; in the recorded morphotype, teeth are increasing in size toward mola. Antenna 5-segmented; basal segment shorter than flagellum; Lauterborn organs absent.

Remarks. The shape of the mentum and mandible are unique for *Nilothauma*, and it is not likely to be confused with other genera.

Ecology and Distribution. Nilothauma is a species-rich genus with worldwide distribution. Larvae are known from littoral and sublittoral soft sediments of standing waters and also form different types of flowing waters. Rare in Central American subfossil material.

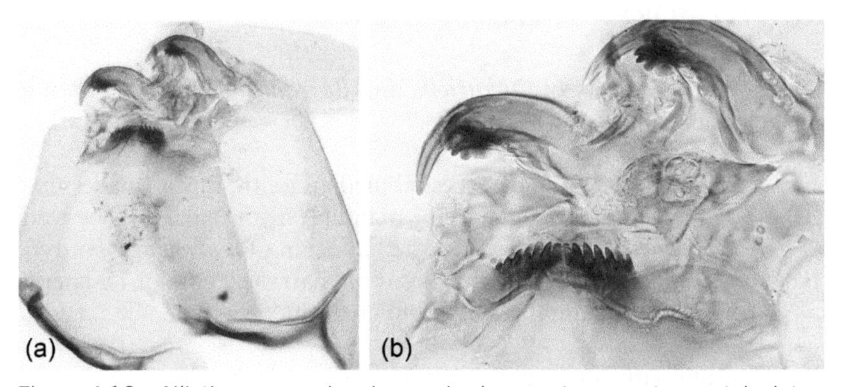

Figure 4.16 *Nilothauma.* a—head capsule, b—mentum, ventromental plates, and mandibles (note the characteristic hooked shape).

OUKURIELLA EPLER

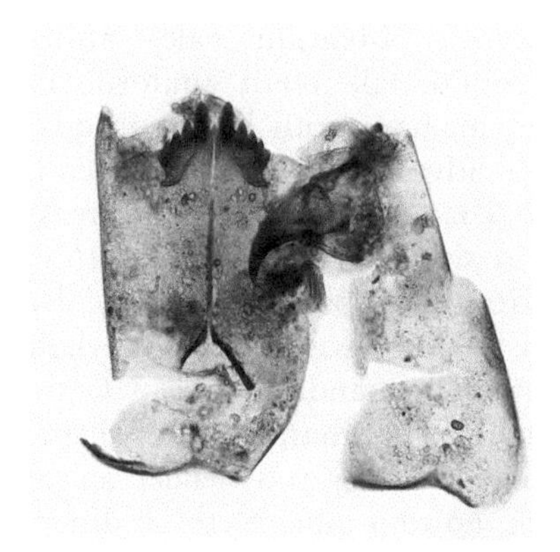

Diagnosis. Mentum with recessed simple or bifid median tooth, pale or dark (Figure 4.17a,c); 5 pairs of lateral teeth decreasing in size laterally, 1st lateral tooth much larger relative to median tooth; ventromental plates short, curved and coarsely striated, medially separated by distance between the first two lateral teeth. Premandible bifid. Mandible with short and dark apical tooth and 3 inner teeth, dorsal tooth present (Figure 4.17b,d). Antenna 6-segmented; Lauterborn organs large, alternately located at the apices of 2nd and 3rd segments.

Remarks. In general, larval *Oukuriella* closely resemble *Beardius*; however, *Oukuriella* has 3 inner teeth on the mandible, *Beardius* only 2. Median mental tooth is always pale in *Beardius*, but can be both pale and dark in *Oukuriella*. These two genera are problematic to separate due to the frequently missing mandible in subfossil material. However, the shape of the lateral end of the dorsoventral plates seems reliable for distinguishing them as subfossils. in *Oukuriella* pointing forward, and in *Beardius* pointing either laterally or slightly backward (L. Pinho, pers. comm.; see also figures in Trivinho-Strixino 2014).

KEY TO MORPHOTYPES

Due to high similarity of *Oukuriella* and *Beardius*, we included both in the key of morphotypes.

1 Mentum with small dark recessed median tooth, much smaller than 1st lateral, with 5 lateral teeth; mandible with 3 dark inner teeth .... *Oukuriella pinhoi*-type

1' Mentum with median tooth pale, about half the length of 1st lateral, which is dark, mandible with 2 inner teeth 2

2 Mentum with small, pale, distinctly triangular median tooth and 4 lateral teeth, 1st lateral dark, broad and noticeably round in shape, mentum strongly arched .... *Beardius* type Chanmico

2' Mentum with pale median tooth, 5 dark lateral teeth mentum not strongly arched .... *Beardius* type Ipala

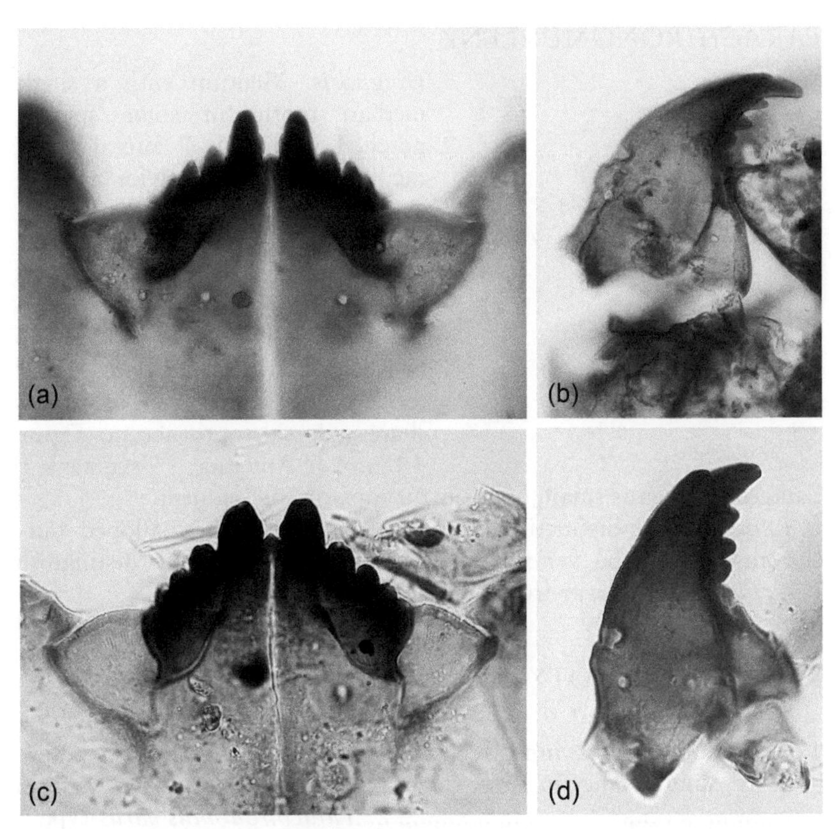

Figure 4.17 *Oukuriella. Oukuriella pinhoi*-type: a—mentum, b—mandible, c—mentum (worn down), d—mandible (worn down); note the direction of the lateral end of ventromental plates.

ADDITIONAL REMARKS TO MORPHOTYPES

Oukuriella pinhoi-type (see Fusari et al. 2014) has unique features, such as a small, dark, recessed median tooth, the same color as the 5 lateral teeth, mandible with 3 dark inner teeth and 1 dorsal tooth.

Ecology and Distribution. Oukuriella is limited to the Neotropical region, where it is widespread, with four named species from Central America (Spies et al. 2009). Larvae are associated with freshwater sponges and immersed wood in stream and rivers (Fusari et al. 2014). Rarely recorded in Central American surface sediments.

PARACHIRONOMUS LENZ

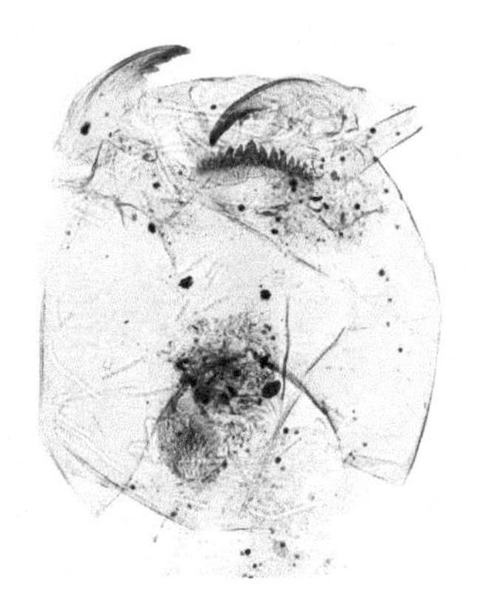

Diagnosis. Mentum with a single median tooth (in some species, notched medially); 7 lateral teeth, median tooth about twice as wide as 1st lateral, outermost tooth minute and depressed (Figure 4.18a,b). Ventromental plates triangular with crenulated to scalloped anterior margin. Premandible with 2–4 teeth, no brush. Mandible with long apical tooth and 2–3 inner teeth of various shapes, dorsal tooth absent (Figure 4.18a,b). Antenna 5-segmented; Lauterborn organs small, opposite on apex of 2nd segment.

Remarks. The distinctive shape of mentum with the scalloped anterior margin of the ventromental plates will sufficiently distinguish *Parachironomus* larvae from all other Chironomini.

KEY TO MORPHOTYPES

1 Mandible with 2 inner teeth; mentum with lateral teeth gradually diminishing in size, forming a concave toothed margin ....*Parachironomus varus*-type A

1' Mandible with 3 inner teeth; mentum with the 4th lateral tooth lower than adjacent teeth, forming a concave toothed margin on each side of the mentum. ....*Parachironomus varus*-type B

ADDITIONAL REMARKS TO MORPHOTYPES

The two morphotypes identified had similar menta (slightly convex edge with a single pointed median tooth, 7 lateral teeth, outermost minute, and paler than other lateral teeth). A morphotype with this type of mentum is referred to as *Parachironomus varus*-type (Brooks et al. 2007) or *Parachironomus arcuatus* group (Epler et al. 2013). However, *Parachironomus varus*-type A (Figure 4.18a) has mandible with 2 inner teeth, and lateral mental teeth gradually diminishing in size, forming a convex toothed margin. *Parachironomus varus*-type B (Figure 4.18b) has 3 inner mandibular teeth, mentum with 4th lateral tooth lower than 3rd and 5th, forming a concave toothed margin on each side of the mentum.

Ecology and Distribution. It is a widespread and species-rich genus, with five named species known from Central America (Spies et al. 2009).

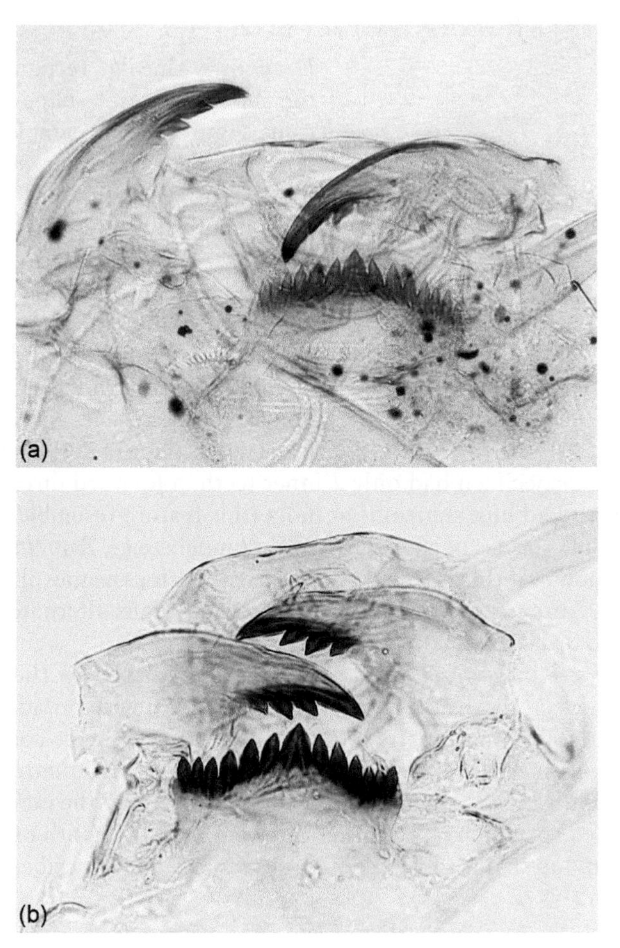

Figure 4.18 *Parachironomus. Parachironomus varus*-type A: a—mentum and mandible; *Parachironomus varus*-type B: b—mentum and mandible.

Larvae of *Parachironomus* live in wide range of habitats of standing and flowing waters. An ecologically versatile group, some species mine in Bryozoa or aquatic plants, or are ectoparasitic on other invertebrates. Common across Central American material.

PARALAUTERBORNIELLA LENZ

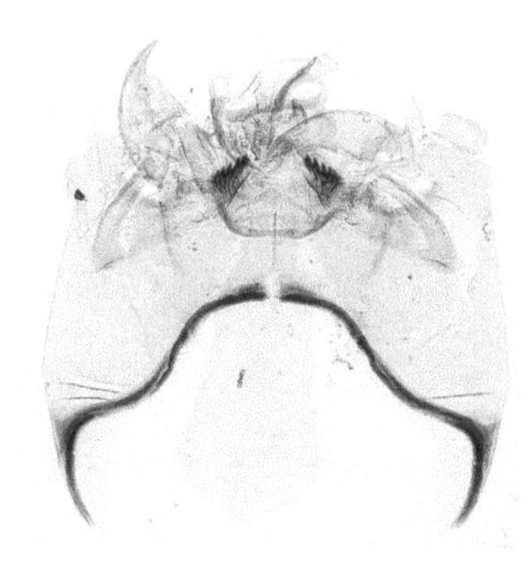

Diagnosis. Easily recognizable by the characteristic shape of mentum (Figure 4.19a). simple, broadly domed, and transparent median tooth, clearly delineated from the dorsomentum with 6 dark lateral teeth evenly decreasing in size laterally. Ventromental plates widely separated medially, long, curved, and coarsely striated. Premandible bifid. Mandible with slender apical tooth, dorsal tooth absent and 3 distinct inner teeth (Figure 4.19b). However, the recorded specimen had only 2 inner teeth, a forward directed tapering tooth and a blunt short one at mola (this feature resembles the mandible of some species of the *Harnischia*-complex, e.g., *Paracladopelma*). It is not known, if this is a characteristic feature for the morphotype or a deformity. Antenna 6-segmented; Lauterborn organs alternate on apices of 2nd and 3rd segments.

Remarks. Paralauterborniella can be distinguished by the mandible without a dorsal tooth and the distinctive mentum with broad, transparent median tooth and 6 pairs of darker lateral teeth, wide coarsely striated ventromental plates. The only described species, *Paralauterborniella nigrohalteralis* (Malloch), has 3 distinct inner teeth on the mandible, but the recorded specimen has a rather different mandible with only 2 inner teeth of unique shape and is likely to represent an undescribed species.

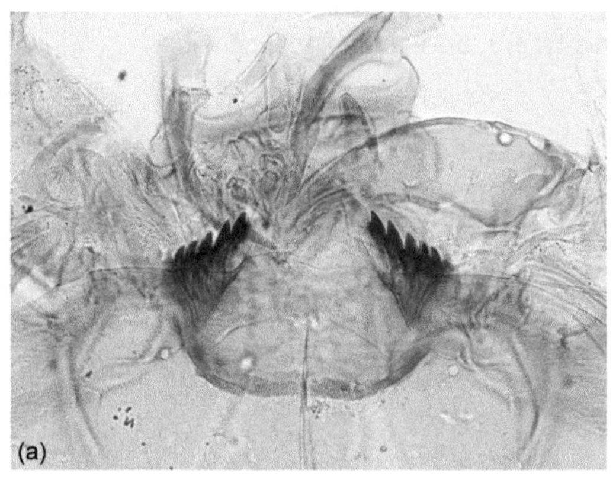

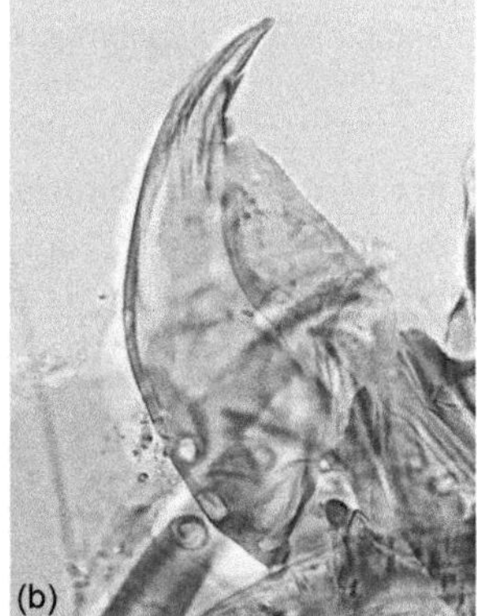

Figure 4.19 *Paralauterborniella.* a—detail of head with mentum and mandible, b—mandible.

Ecology and Distribution. Paralauterborniella nigrohalteralis is a species with broad distribution pattern and known also from Costa Rica and Nicaragua (Spies and Reiss 1996). Larvae inhabit lakes (littoral), old river branches, and slow flowing rivers (Moller Pillot 2009). Rare in Central American lakes.

PARATENDIPES KIEFFER

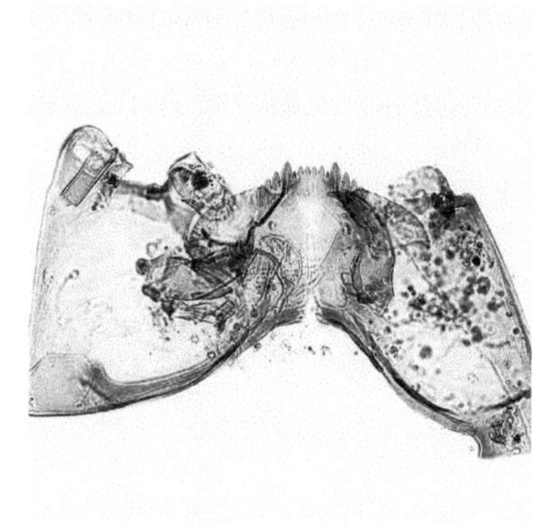

Diagnosis. Mentum with 4 uniform pale median teeth, and 6 lateral teeth; 1st and 2nd lateral fused, 2nd lateral higher than median teeth; remaining lateral teeth decreasing gradually (Figure 4.20a). Ventromental plates curved with coarse striation in the anterior half; ventromental plate delineation associated with the base of 2nd (in some species, 1st) pair of lateral teeth. Premandible with 2–3 teeth and a basal tooth. Mandible with 2 inner teeth and 1 dorsal tooth; pigmentation of mola may resemble 3rd inner tooth (Figure 4.20b). Antenna 6-segmented; Lauterborn organs alternate on 2nd and 3rd segments.

Remarks. Paratendipes larvae can be distinguished by the 4 pale median mental teeth, which are always lower than the 2nd lateral tooth (in addition to the 6-segmented antenna with alternate Lauterborn organs). A single morphotype was recorded in the study material, *Paratendipes nudisquama*-type following Pinder and Reiss (1983).

Ecology and Distribution. Larvae of *Paratendipes* occur in broad variety of standing and flowing waters; they prefer soft sediments and sandy bottoms. Rare in Central American lakes.

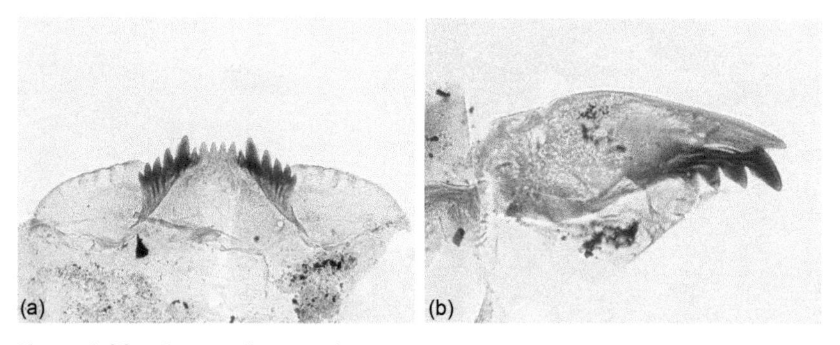

Figure 4.20 *Paratendipes nudisquama*-type. a—mentum, b—mandible.

PELOMUS REISS

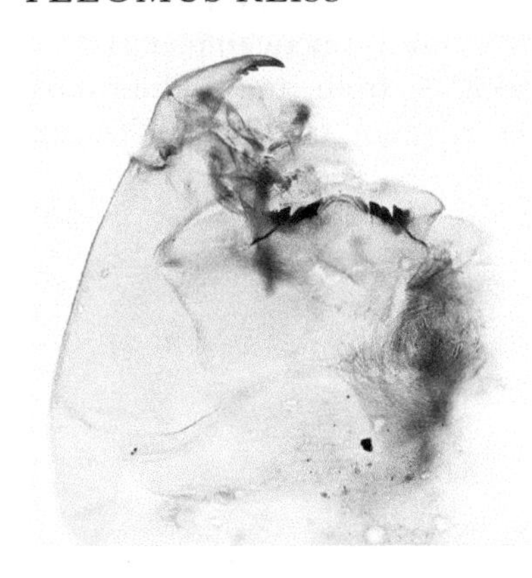

Diagnosis. Mentum with 4 wide light median teeth, slightly convex, followed by 3 pairs of darker teeth and several fine pale teeth (Figure 4.21a,b,d); there is a large gap between 2nd and 3rd lateral teeth. Wide ventromental plates with coarse striation and smooth anterior margin. Premandible with 2 large teeth and 1 small tooth. Mandible with long apical tooth and 2 inner teeth, dorsal tooth absent (Figure 4.21c). Antenna 6-segmented; Lauterborn organs not observed (Trivinho-Strixino and Strixino 2008).

Remarks. The genus resembles *Saetheria* and *Cryptochironomus*. However, the mentum with 4 median teeth, as opposed to a single broad median tooth in the other two genera and with its distinct shape and color pattern are distinctive for *Pelomus*.

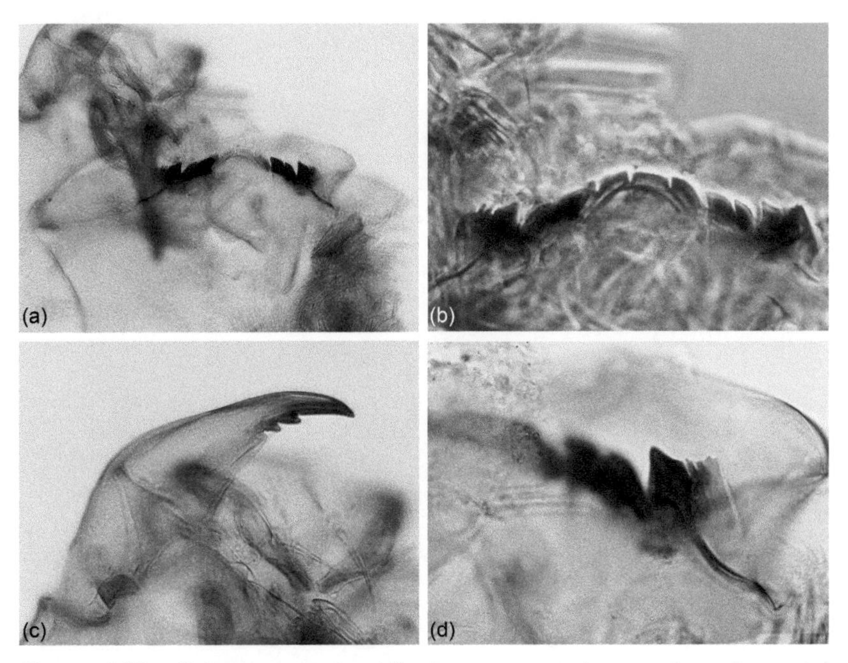

Figure 4.21 *Pelomus psammophilus*-type. a—mentum and ventromental plates, b—detail of mentum, c—mandible, d—detail of outer lateral teeth of mentum.

Pelomus psammophilus-type (after Trivinho-Strixino and Strixino 2008) was the only type identified in the Central American material.

Ecology and Distribution. Larvae are known from sandy sediments of small reservoirs and littoral of big lakes (Trivinho-Strixino and Strixino 2008). Rare in Central American lakes.

POLYPEDILUM KIEFFER

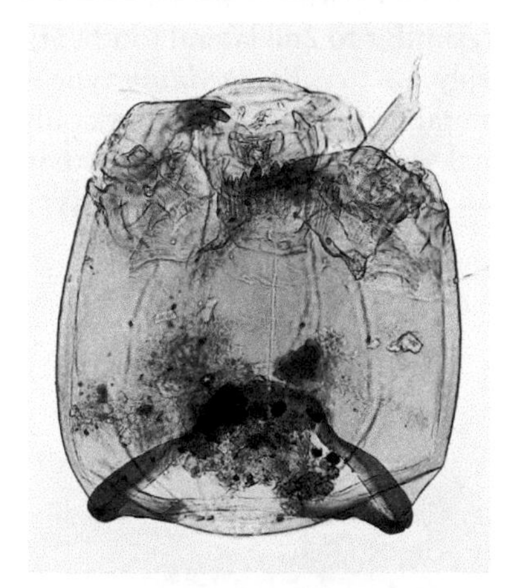

Diagnosis. Pattern of mental teeth varies, but usually with double median tooth and 7 (Figure 4.22a,d); 1st lateral tooth may be minute (in this case, 2nd lateral subequal to median teeth; Figure 4.22d–f) or subequal to other teeth (in this case, all mental teeth of similar size; Figure 4.22a–c), outermost tooth short and recessed. Ventromental plates variable but broad and widely separated medially (by at least the width of median teeth); striae continuous. Premandible with 3 teeth (2 apical and one blunt basal tooth). Mandible with all teeth dark, with 2–3 inner teeth, dorsal tooth usually present, occasionally absent (Figure 4.22b–d). Antenna 5-segmented; Lauterborn organs opposite on apex of 2nd segment. Relative lengths of antennal segments 2–5 is an important identification character, e.g., 3rd segment is minute in some species of the subgenus *Tripodura*.

Remarks. The distinctive mentum, with median and 2nd lateral teeth longer than 1st one, or the 4 median teeth subequal in size to lateral ones distinguishes most members of the genus. Moreover, the apically widened frons is typical for most larval types.

KEY TO MORPHOTYPES

1 Mentum with median and lateral teeth subequal in size.................2
1' Mentum with teeth of various sizes: median teeth extended, 1st lateral teeth minute and 2nd lateral subequal to median teeth............3

2 Ventromental plates wide and slender, longer than width of mentum, dorsal tooth of mandible large, subequal to apical tooth ....*Polypedilum beckae*-type
2' Ventromental plates narrower, fan-like, dorsal tooth of mandible of regular size, smaller than apical tooth........*Polypedilum fallax*-type

3 Lateral teeth diminishing in size gradually, 4th lateral tooth not smaller than 5th lateral tooth.........*Polypedilum nubeculosum*-type
3' 4th lateral tooth shorter than either of the adjacent teeth...............4

4 Outermost lateral teeth forming a distinct pattern: 4th lateral very small, 5th much longer and broader, similar to 2nd lateral tooth, 5th to 7th teeth diminishing in size steeply..............*Polypedilum* type A

4' No such pattern in the outermost lateral teeth, 5th lateral tooth only slightly higher than 4th; 3rd antennal segment minute, shorter than 4th..*Polypedilum tripodura*-type

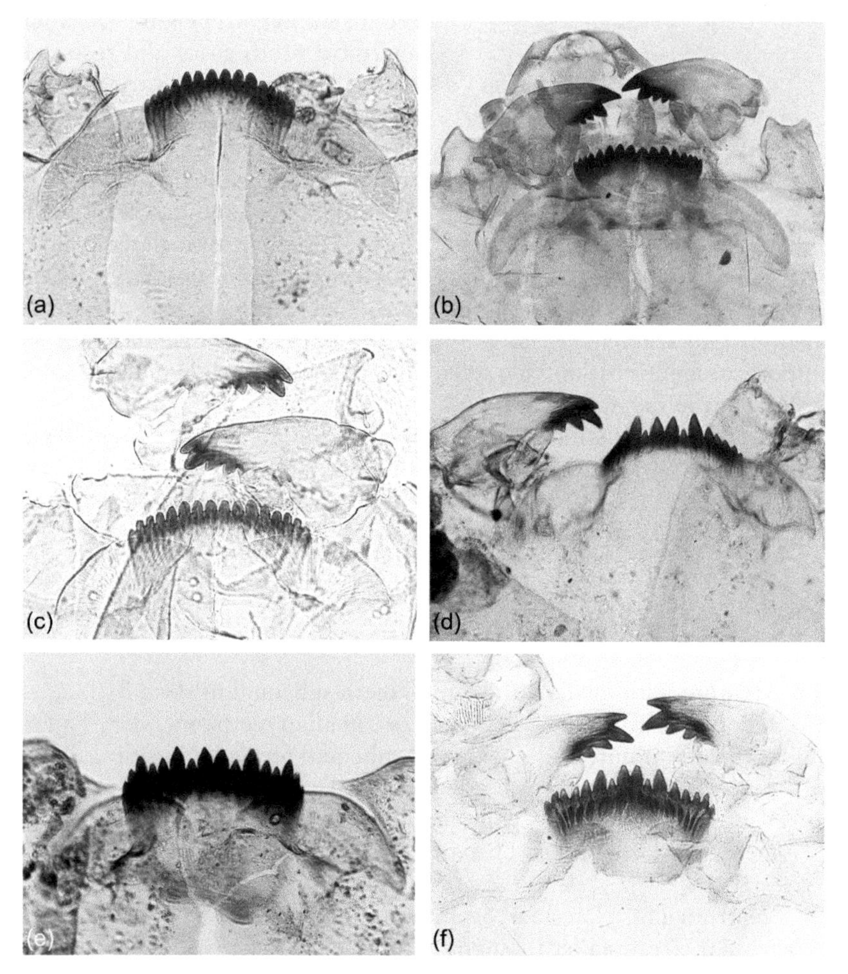

Figure 4.22 *Polypedilum. Polypedilum beckae*-type: a—mentum, b—mentum and mandible; *Polypedilum fallax*-type: c—mentum and mandible; *Polypedilum nubeculosum*-type: d—mentum and mandible; *Polypedilum* type A: e—mentum; *Polypedilum tripodura*-type: f—mentum and mandible.

ADDITIONAL REMARKS TO MORPHOTYPES

Five morphotypes were identified.

Polypedilum beckae-type (Figure 4.22a,b). Mandible with 2 inner teeth, dorsal tooth equal to or larger than apical tooth. Ventromental plates similar to the ones in the subgenus *Asheum*, slender and longer than width of mentum.

Polypedilum fallax-type (Figure 4.22c). Mandible with 2 inner teeth and dorsal tooth smaller than apical tooth. Ventromental plates fan-like.

The three remaining morphotypes have typical "*Polypedilum*-like" menta, i.e., with extended median teeth, minute 1st lateral and 2nd lateral teeth subequal to median ones in size, and mandible with 2 inner teeth.

Polypedilum nubeculosum-type (Figure 4.22d). It can be differed from the previous two types in the arrangement of the lateral teeth that gradually diminish laterally in size.

Polypedilum type A and *Polypedilum tripodura*-type. Menta similar, with 4th lateral tooth shorter than the two adjacent ones. These types can be distinguished based on the arrangement of the outermost 3 lateral teeth: *Polypedilum* type A (Figure 4.22e) has a distinct cluster decreasing in size, looking like a steep "slope", mostly like *Polypedilum* sp. 2 and sp. 6 (after Trivinho-Strixino 2014), while in *Polypedilum tripodura*-type (Figure 4.22f) the 6th lateral tooth is only slightly lower than the penultimate one. The antenna of the latter is similar to that one of the subgenus *Tripodura*, with a minute 3rd antennal segment.

Ecology and Distribution: This is a large genus with worldwide distribution, including hundreds of species. Larvae live virtually in all kinds of still and flowing waters where they usually prefer soft sediments, but also occur in coarse habitats or mine in aquatic vegetation. In paleolimnological reconstructions *Polypedilum* remains are considered as indicators of temperate climatic conditions (Brooks et al. 2007). Common in lakes across Central America.

STENOCHIRONOMUS KIEFFER

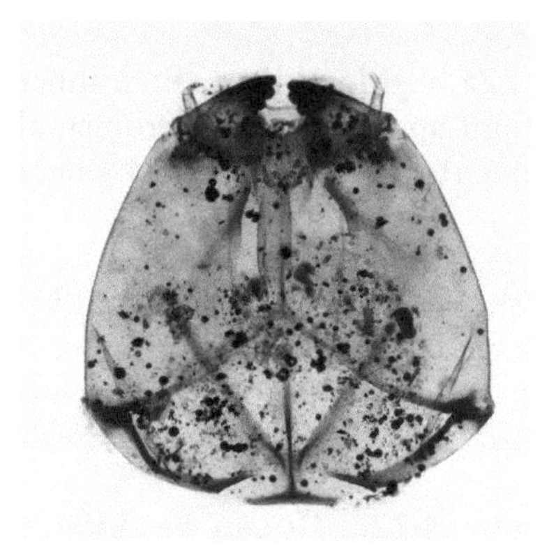

Diagnosis. Shape of head capsule is characteristic, dorsoventrally flattened, wedge-shaped. Mentum distinctive, concave, with 10–20 strongly pigmented teeth; 4 uniform median teeth and 3–4 pairs of elevated lateral teeth (Figure 4.23a). Ventromental plate dramatically reduced forming a large weakly striated lobe (Figure 4.23a). Premandible with 3 teeth. Mandible triangular with the proximal half-black (Figure 4.23b); apical tooth and 2 inner teeth present, dorsal tooth absent, instead, there is a ridge with 2 teeth, which is usually not visible in the subfossils. Antenna 5-segmented; Lauterborn organs minute, opposite on apex of 2nd segment.

Remarks. Stenochironomus has a unique shape of head and mentum but can be confused with *Xestochironomus* (and *Harrisius* Kieffer, known only from Australia). The number of teeth on mentum is indicative for both genera. *Xestochironomus* has 8 mental teeth, while *Stenochironomus* has 10–12. The morphotype found in Central America has 10 teeth.

Ecology and Distribution. Larvae of *Stenochironomus* mine both in living and dead vegetation, including wood in mesotrophic and eutrophic lakes. Due to the specific biology, the genus is rare in Central American lake sediments.

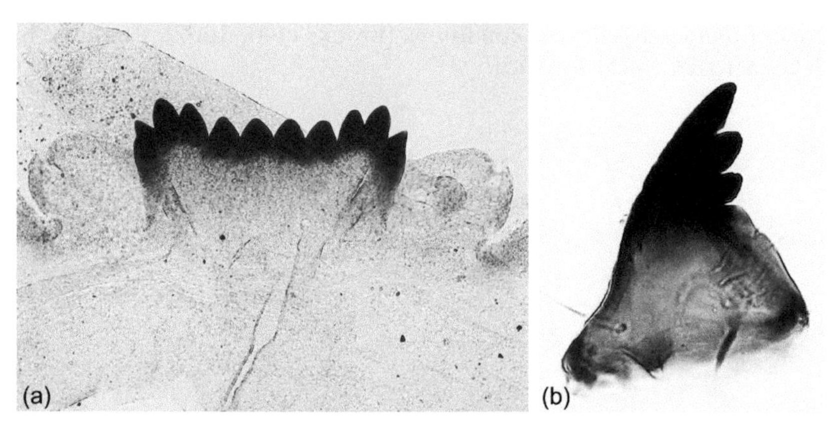

Figure 4.23 *Stenochironomus*. a—mentum, b—mandible.

XENOCHIRONOMUS KIEFFER

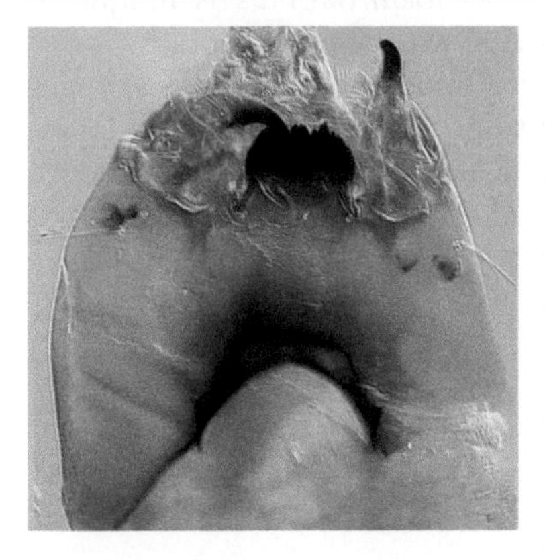

Diagnosis. Strongly pigmented mentum with recessed trifid median tooth with high, domed central part and 2 minute outer teeth (Figure 4.24); 7 lateral teeth, 1st broad and high, higher than median tooth, 2nd narrow and low, 3rd higher than 2nd, remainder decreasingly gradually. Ventromental plates broad and deep, separated by the width of the median teeth, striae fine and continuous (Figure 4.24). Premandible without teeth. Mandible with a long apical tooth and 3 minute inner teeth, dorsal tooth absent. Antenna 5-segmented; Lauterborn organs opposite on apex of 2nd segment.

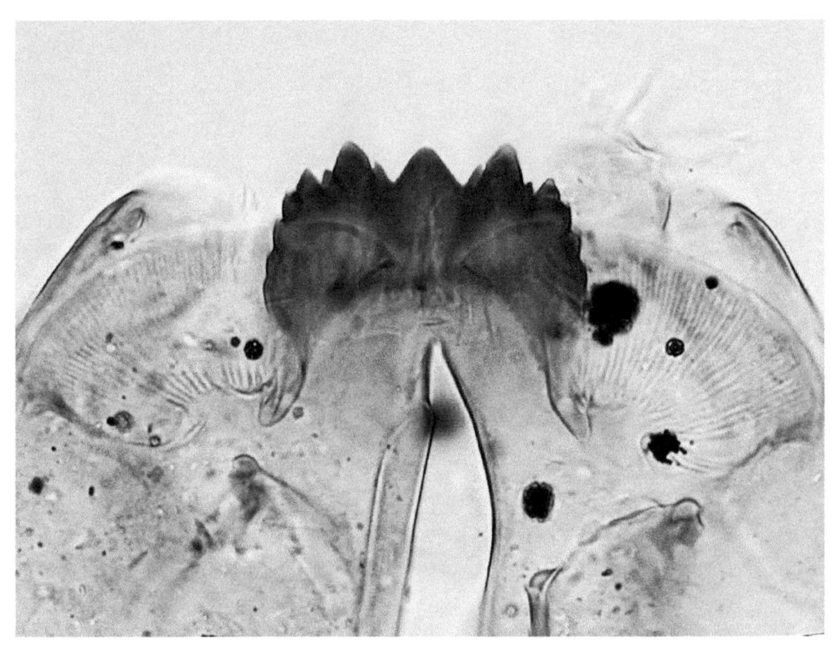

Figure 4.24 *Xenochironomus.* Mentum and ventomental plates.

Remarks. The shape of mentum of *Xenochironomus* is unique in Chironominae and unlikely to be confused with other genera.

Ecology and Distribution. Larvae are miners in freshwater sponges in standing and flowing waters. As a result of their feeding habits, larval mouth parts are frequently worn down. Rare in Central American lakes.

XESTOCHIRONOMUS SUBLETTE & WIRTH

Diagnosis. Head capsule dorsoventrally flattened with a Y-shaped dorsal design Mentum strongly sclerotized, concave, with 8 teeth, most lateral teeth continuous with edge of ventromental plate (Figure 4.25). Ventromental plates vestigial without striation. Mandible robust, triangular apically, with 3–4 distinct teeth; apical tooth short, without dorsal tooth (Figure 4.25). Antenna 5-segmented; Lauterborn organs absent. Premandible with uncertain number of teeth.

Remarks. Xestochironomus has a unique shape of head and mentum and can be confused with *Stenochironomus* (and *Harrisius*, known only from Australia). *Xestochironomus* can be distinguished by having only 8 mental teeth and the unique squared-off shape of the head.

Figure 4.25 *Xestochironomus.* Mentum and mandible.

Ecology and Distribution. Xestochironomus is a species-rich genus, known only from the New World (Andersen and Kristoffersen 1998). Larvae mine in wood in running waters (Borkent 1984). Due to the specific biology of larvae, the genus is rare in Central American lake sediments.

ZAVRELIELLA KIEFFER

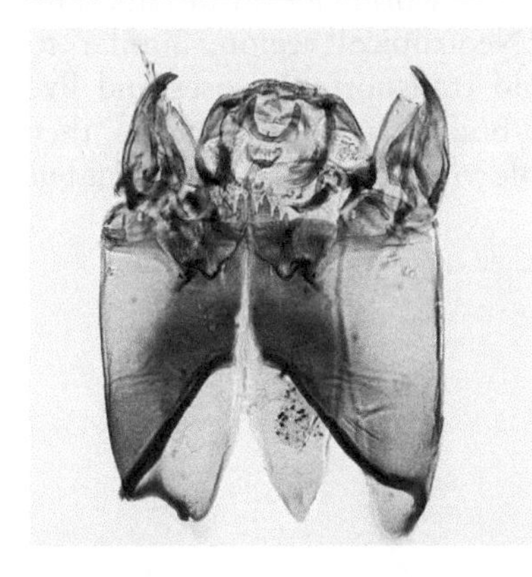

Diagnosis. Mentum with bifid median tooth (Figure 4.26a) and 6 pairs of lateral teeth; 1st lateral small, partially fused with 2nd lateral tooth that is as tall as median teeth. Ventromental plates nearly triangular, about as wide as mentum, almost touching medially, anterior margin smooth, striae restricted to basal and lateral part (Figure 4.26a). Premandible with 3 apical teeth, and a single blunt basal tooth (Figure 4.26a). Mandibular teeth pale brown; apical tooth relatively short, 2 inner teeth and a strong dorsal tooth present. Antenna 6-segmented, 4th segment longer than 3rd and 5th (Figure 4.26b); Lauterborn organs large and alternate on 2nd and 3rd segments.

Remarks. The genus is similar to *Lauterborniella* from which it can be distinguished by the partially fused 1st and 2nd lateral teeth of mentum that are distinctly separate in *Lauterborniella.*

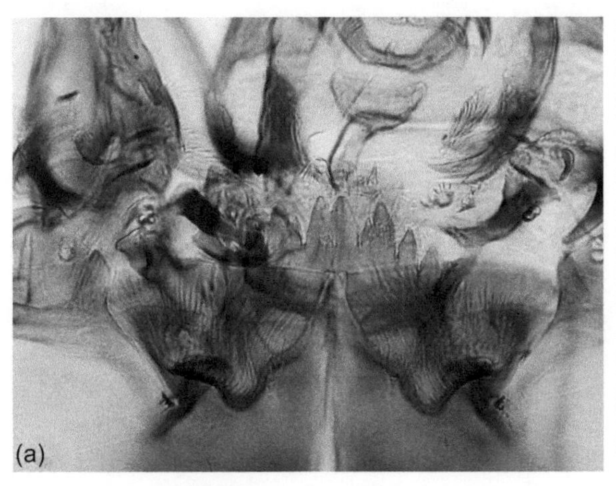

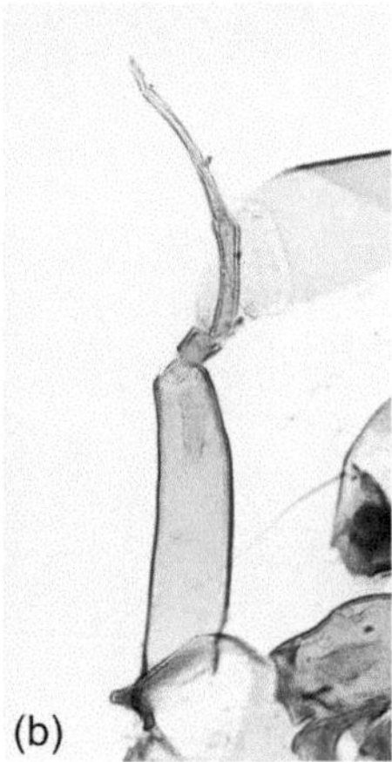

Figure 4.26 *Zavreliella.* a—detail of head showing mentum, ventromental plates, and premandible, b—antenna.

Ecology and Distribution. The genus is worldwide in distribution, with several species reported to the Neotropical region. Similar to *Lauterborniella*, larval *Zavreliella* build transportable cases and live among aquatic vegetation in small waterbodies. In tropical realms, they also dwell in sediments in standing and flowing waters. Not frequent but present across Central America.

TRIBE PSEUDOCHIRONOMINI

Ventromental plates bar-like, in near contact medial. Antenna not mounted on distinct pedestals; Lauterborn organs not placed on pedicels. Seta subdentalis on the same side of mandible as seta interna (dorsal).

KEY TO MORPHOTYPES OF PSEUDOCHIRONOMINI

1 Mentum with 2nd lateral tooth minute, shorter than 1st and 3rd teeth 2

1' Mentum with 2nd lateral tooth longer than 1st and 3rd lateral teeth, 1st lateral tooth shorter than other lateral teeth ....Pseudochironomini type A

2 Mentum of 11–13 teeth of more arcuate shape, with the outermost 1–2 lateral teeth vestigial or fused ....*Pseudochironomus* type Las Pozas

2' Mentum of horizontal shape bearing 13 teeth, outermost lateral teeth separate .... *Pseudochironomus prasinatus*-type

PSEUDOCHIRONOMUS MALLOCH

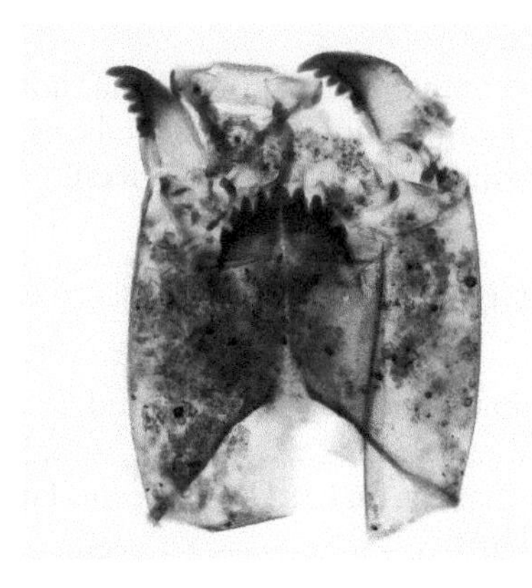

Diagnosis. Mentum with simple, broad, rounded median tooth, about as high as 1st lateral, with 6 lateral teeth, 2nd lateral greatly reduced (in worn down menta, appears fused to the 1st lateral), remaining 4 teeth decreasing progressively, outermost may be fused (Figure 4.27a,c,d). Ventromental plates Tanytarsini-like (Figure 4.27c,d), slender and scarcely separated medially, with a characteristic stalk-like structure protruding below the plate. Mandible with pale apical tooth and 4 dark inner teeth, basal one blunt, dorsal tooth absent (Figure 4.27b,f). Antenna 5-segmented; Lauterborn organs small, opposite on 2nd segment (Figure 4.27b).

Remarks. Larvae of *Pseudochironomus* resemble those of Tanytarsini due to the slender ventromental plates, which are almost in contact medially and has a narrow central band of fine striae. However, the genus can be distinguished by the antenna not mounted on distinct pedestals and the Lauterborn organs not placed on pedicels.

ADDITIONAL REMARKS TO MORPHOTYPES

Two morphotypes were identified in the Central American material (see key on page 85) following Epler (2001) who distinguish them as two major *Pseudochironomus* types, without association to morphotypes or species groups.

Pseudochironomus Las Pozas-type (Figure 4.27a–c) has a characteristic arched mentum with 11–13 teeth; 1st and 2nd lateral teeth are fused, last 1–2 teeth are vestigial or fused to a lobe at the base of mentum.

Pseudochironomus prasinatus-type (Figure 4.27d–f) possesses mentum of linear shape bearing 13 teeth, outermost lateral teeth separate. *Pseudochironomus prasinatus*-type most resembles *Tanytarsus* and *Cladotanytarsus mancus*-type. However, it can be distinguished from both by the specific shape of ventromental plates with the stalk-like extension on the posterior margin and by 4 lateral teeth outside the minute lateral tooth, in contrast to only 3 in *Cladotanytarsus mancus*-type.

Ecology and Distribution. Larvae prefer sandy substrata of lake littorals and rivers; they may also be found in brackish/estuarine water (Epler 2001). Rare in Central American lakes.

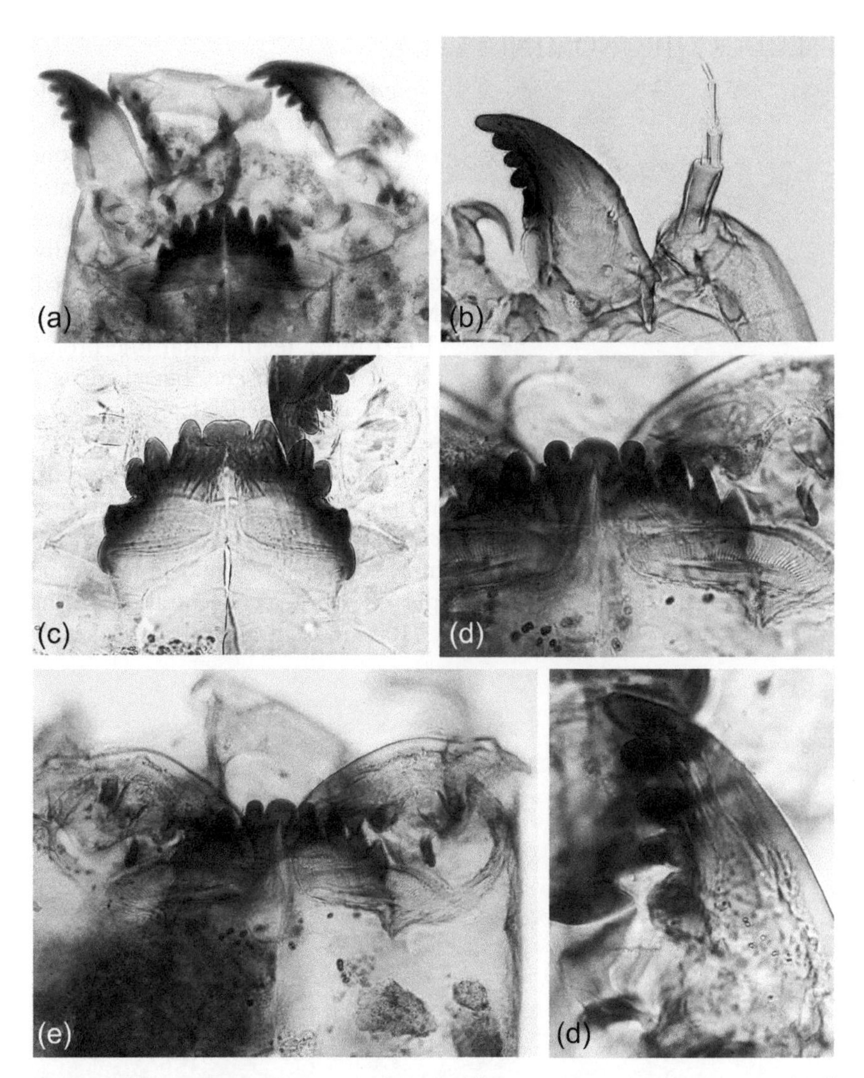

Figure 4.27 *Pseudochironomus. Pseudochironomus* type Las Pozas: a—detail of head, b—mandible and antenna, c—mentum; *Pseudochironomus prasinatus*-type: d—mentum, e—detail of head, f—mandible.

PSEUDOCHIRONOMINI TYPE A

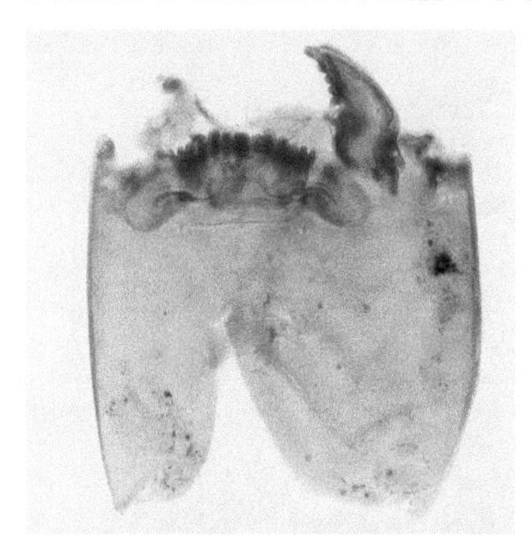

Diagnosis. Mentum has a single median tooth and 6 lateral teeth, 1st lateral tooth smaller than 2nd one (Figure 4.28a). Ventromental plates short, "Tanytarsini-like" (Figure 4.28a), almost connecting medially, anterior margin strongly curved. Mandible with pale apical tooth and 3 (4?) inner teeth, basal one broad, dorsal tooth absent, but there is a ridge with 1 surficial tooth (Figure 4.28b).

Remarks. This morphotype most likely belongs to the tribe Pseudochironomini but its generic position is uncertain; it may belong to *Pseudochironomus*, *Riethia*, or *Manoa*.

Ecology and Distribution. The morphotype was found in a lowland waterbody with temperature 25°C, pH 7.8, oxygen concentration 7.3 mg L^{-1}, and conductivity of 2040 µS cm^{-1}. Rare in Central American lakes.

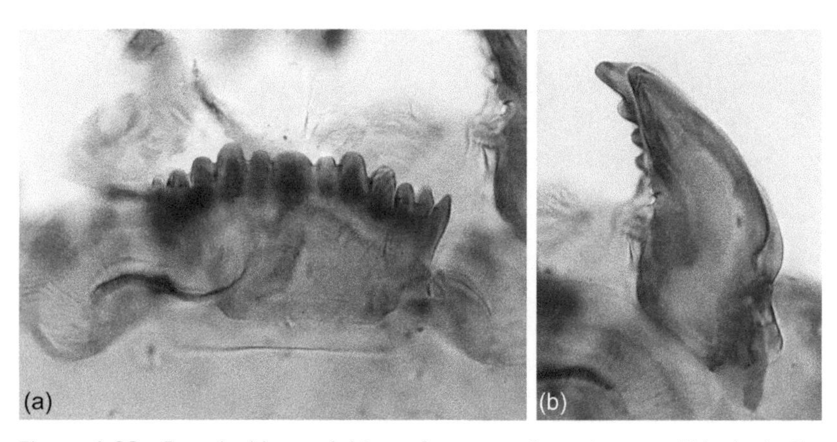

Figure 4.28 Pseudochironomini-type A. a—mentum, b—mandible (note the ridge with a surficial tooth).

TRIBE TANYTARSINI

Antenna mounted on distinct elongated pedestals; Lauterborn organs usually well developed and often situated on short to long pedicels. Mentum with single median tooth (may be laterally notched) and mostly with 5 lateral teeth. Ventromental plates usually wide and slender, almost touching medially, in some genera narrower, separated medially by at least the width of the three median teeth. Figure 4.29 shows the Tanytarsini larval head capsule with typical subfossil features.

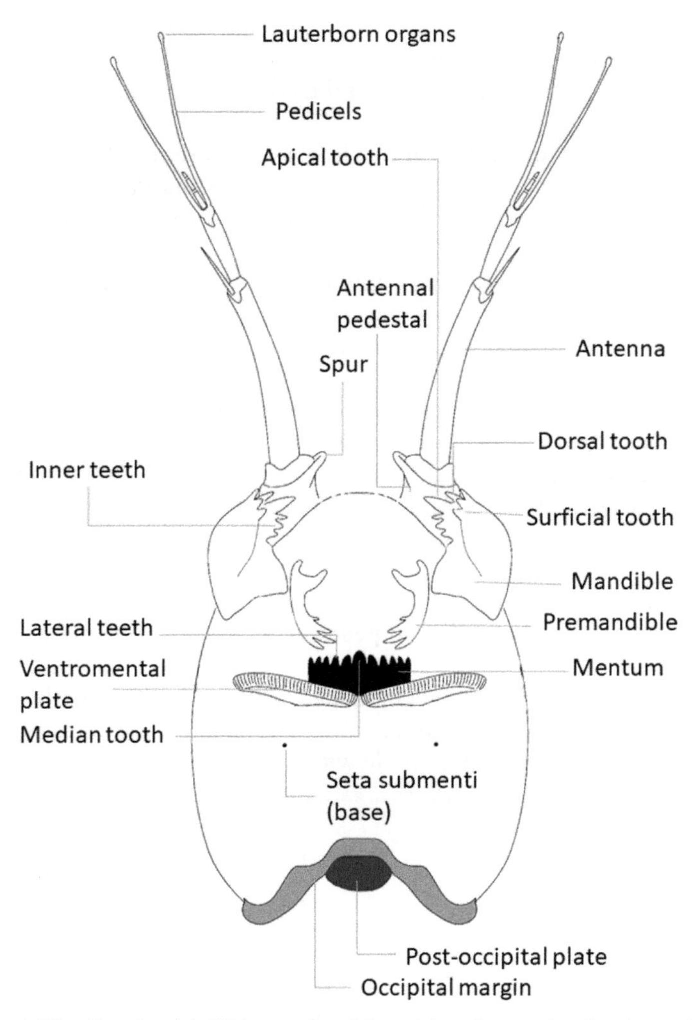

Figure 4.29 Tanytarsini (Chironominae) larval head capsule showing subfossil characters.

KEY TO GENERA OF TANYTARSINI*

1 Mentum with 4 lateral teeth............................. Tanytarsini type A
1' Mentum with 5–6 lateral teeth 2

2(1) Large multi-branched spur on antennal pedestal......... *Stempellina*
2' No such spur on antennal pedestal or, if spur present, it is simple 3

3(2) Mentum with 2nd lateral tooth minute, shorter than 1st and 3rd teeth.. *Cladotanytarsus* in part
3' Mentum with 2nd lateral tooth subequal to 1st and 3rd teeth or 1st lateral shorter than other lateral teeth... 4

4(3) First lateral tooth of mentum clearly shorter than other lateral teeth .... *Cladotanytarsus* in part
4' First lateral tooth of mentum subequal to 2nd lateral tooth......... 5

5(4) Antennal pedestal short, about as long as broad 6
5' Antennal pedestal prolonged, about twice as long as broad 7

6(5) Ventromental plates of characteristic shape: strongly curved at lateral end and coarsely striated; median mental tooth laterally notched .... *Rheotanytarsus*
6' Ventromental plates straight, without coarse striae, median mental tooth rounded, at most weakly notched laterally ...... *Paratanytarsus*

7(5) Spur present on antennal pedestal.. 8
7' Spur absent from antennal pedestal................... *Tanytarsus* in part

8(7) Head capsule strongly pigmented, dark, mandible with 3 inner teeth, premandible bifid... *Micropsectra*
8' Head capsule pale; mandible with 2 inner teeth; premandible with 3 teeth... *Tanytarsus* in part

*Modified after Brooks et al. (2007) and Epler et al. (2013).

CLADOTANYTARSUS KIEFFER

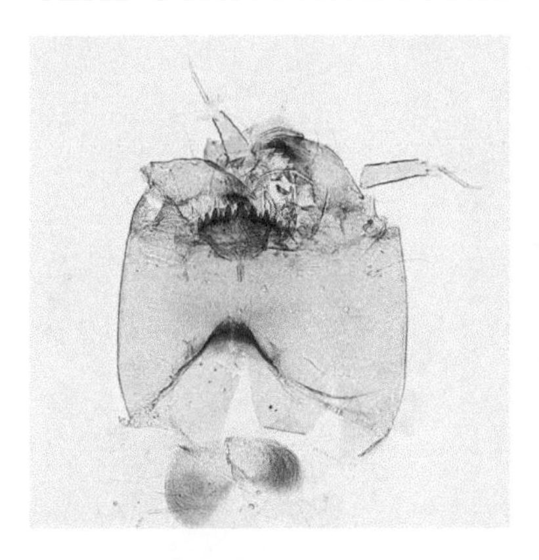

Diagnosis. Mentum with broad median tooth, often notched with outer teeth variable in size and degree of fusion, 5 pair of lateral teeth, regularly decreasing in size or 2nd lateral much smaller than remainder (in *C. mancus*-type, Figure 4.30a). Ventromental plates a little wider than mentum, close medially. Post-occipital plate well developed. Premandible with 4 or 5 apical teeth. Mandible with dark apical tooth and 3 inner teeth, dorsal tooth pale. Antenna characteristic, with 2nd segment short, wedge-shaped, 3rd segment longer than 2nd; Lauterborn organs large, opposite on apex of wedge-shaped 2nd segment (Figure 4.30b).

Remarks. The most characteristic feature of the genus is the large Lauterborn organs on short pedicels opposite on the wedge shaped 2nd antennal segment. In addition, the wide ventromental plates is also distinctive to the group.

KEY TO MORPHOTYPES

1 Mentum with 2nd pair of lateral teeth notably smaller than the adjacent lateral teeth.............................. *Cladotanytarsus mancus*-type

1' Mentum with 1st pair of lateral teeth very small, close to the laterally notched median tooth, mentum seems like composed of a compound median tooth and 4 pairs of lateral teeth ..... *Cladotanytarsus* type A

ADDITIONAL REMARKS TO MORPHOTYPES

Two morphotypes were identified following Pinder and Reiss (1983).

Cladotanytarsus mancus-type (Figure 4.30a). It can be easily recognized by the mentum with small 2nd pair of lateral teeth.

Cladotanytarsus type A (Figure 4.30c). It has a distinctive mentum with laterally deeply notched median tooth and small 1st pair of lateral teeth appressed to median tooth, so mentum seems like composed of a compound median tooth and 4 pairs of separate lateral teeth (Pinder and Reiss 1983). Since antenna was missing, the placement of this morphotype to this genus is tentative.

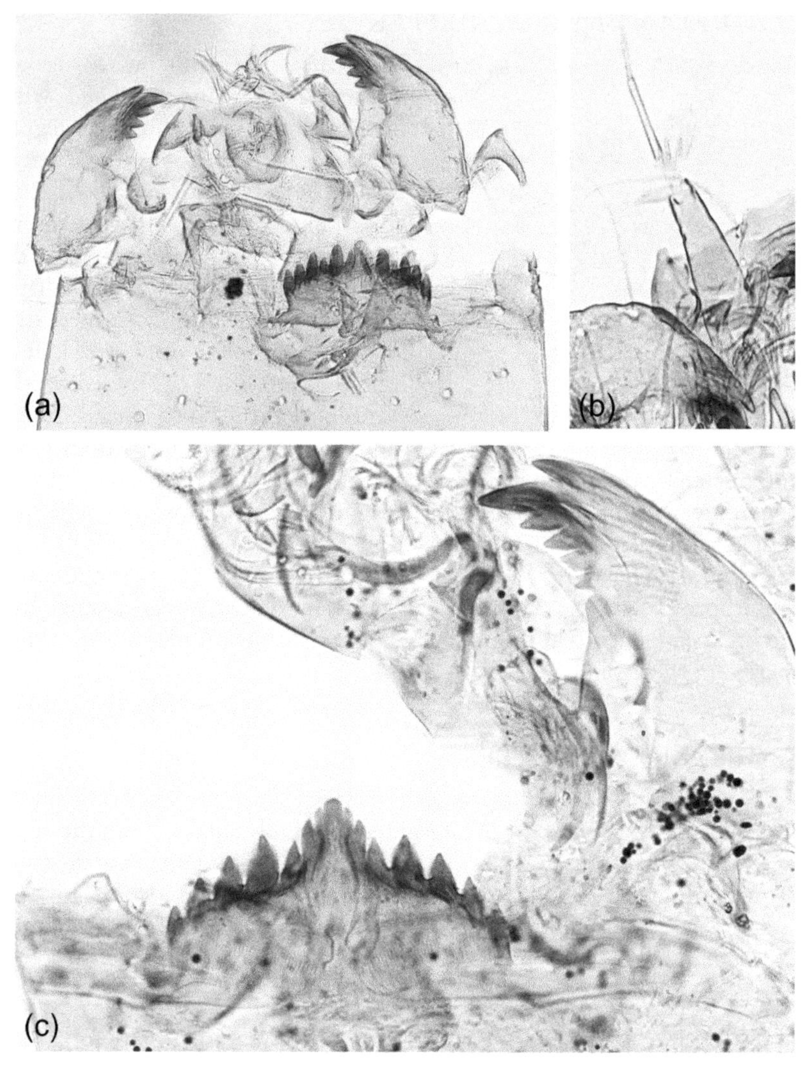

Figure 4.30 *Cladotanytarsus. Cladotanytarsus mancus*-type: a—detail of head with mentum with reduced 2nd lateral teeth, ventromental plates, mandible, and premandible, b—antenna with large Lauterborn organs; *Cladotanytarsus* type A: c—detail of head with mentum, mandible, and premandible.

Ecology and Distribution. Cladotanytarsus is a common and species-rich genus in the Nearctic region (Spies and Reiss 1996). Larvae are ecologically versatile, living in a variety of lentic and lotic habitats, including brackish waters and hot springs. Relatively frequent across Central America.

MICROPSECTRA KIEFFER

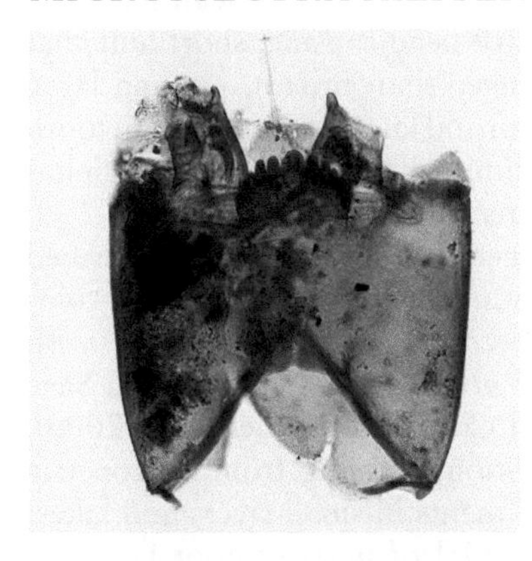

Diagnosis. Mentum with median tooth rounded, laterally notched to crenate, usually pale (Figure 4.31a,b). Ventromental plates close medially, a little wider than mentum, with fine striation. Post-occipital plate present, usually well developed (Figure 4.31c). Premandible bifid. Mandible with dark apical tooth, 3 inner teeth and dorsal tooth. Antenna 5-segmented, located on a pedestal usually with an apparent spur (Figure 4.31b); Lauterborn organs small on long pedicels placed opposite on apex of 2nd segment.

Remarks. Micropsectra most resembles *Tanytarsus*, and the best character separating them is the premandible, which is bifid in *Micropsectra*, while trifid in *Tanytarsus*. In case of missing premandible, the presence of spur on the antennal pedestal has been used to characterize *Micropsectra*. Although some *Tanytarsus* species also possess a spur, according to Brooks et al. (2007), the shape of spur can be distinctive, since it is sharp in *Micropsectra*, but with different shape when compared to *Tanytarsus*. That said, there also are several extant common *Tanytarsus* species in which the spur is quite sharp and long (J. Epler, pers. comm.). *Micropsectra* can be confused with *Paratanytarsus* as well, but while the mentum of *Paratanytarsus* is uniformly dark, in *Micropsectra*, the middle part of mentum is pale.

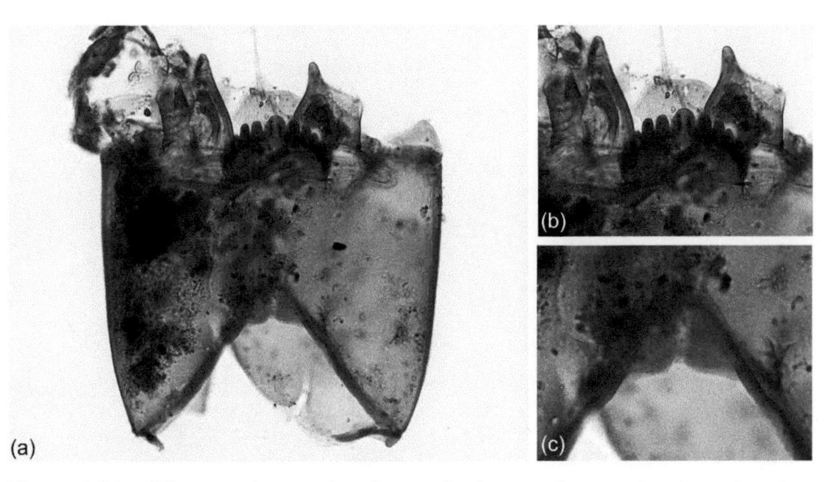

Figure 4.31 *Micropsectra.* a—head capsule, b—mentum and antennal pedestal with spur, c—post-occipital plate.

The only morphotype recognized in the Central American material, *Micropsectra* type Magdalena, has a dark head capsule, short antennal pedestals (about as long or slightly longer than broad), bearing large spur; mentum with dark, rounded median and lateral teeth; median tooth with a pale band, lateral teeth slightly diminishing in size; post-occipital arch narrow, plate present, slightly lighter than occipital margin.

Ecology and Distribution. A large genus with about 100 described species, with fewer known as larvae. Even though *Micropsectra* is common in the Holarctic, it is only rarely recorded in the Neotropics, with only one named species, *Micropsectra atitlanensis* (Sublette and Sasa 1994), and some unnamed adults from Costa Rica (Spies et al. 2009). Larvae are found in a variety of conditions ranging from hygropetric habitats or temporary pools to soft sediments in small rivers and lakes; their typical habitat is littoral to profundal of nutrient-poor lakes. In Central American material, the genus seems to be restricted to large lakes, and lakes located in high elevations. Relatively frequent across Central America.

PARATANYTARSUS THIENEMANN & BAUSE

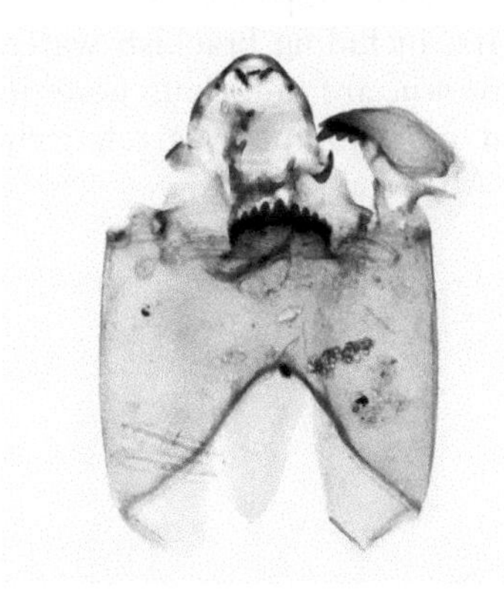

Diagnosis. Median tooth rounded or weakly notched laterally, with 5 pairs of lateral teeth, decreasing in size laterally; all median teeth uniformly dark (Figure 4.32b). Ventromental plates close medially, each plate slightly longer than width of mentum. Post-occipital arch is deeply incised, plate absent (Figure 4.32a). Premandible bifid. Mandible with apical tooth and 2–3 inner teeth similar in color, dorsal tooth brown or yellow (Figure 4.32c). Antenna 5-segmented, placed on short (about as long as broad) pedestal without apical projection (Figure 4.32a); Lauterborn organs small, sessile or placed on very short pedicels opposite on apex of elongate 2nd segment.

Remarks. Subfossil *Paratanytarsus* resembles *Micropsectra* and *Tanytarsus* but can be distinguished by the combination of the shape of post-occipital arch, which is deeply incised, and short antennal pedestal. *Paratanytarsus* also resembles *Neozavrelia*, which has not been recorded in Central America, but it can be distinguished by 5 pairs of well-developed lateral teeth, as opposed to 4 lateral teeth in *Neozavrelia*.

One morphotype was recorded in the Central American material, *Paratanytarsus penicillatus*-type (after Brooks et al. 2007), with 2 inner mandibular teeth.

Ecology and Distribution. This is a species-rich genus with worldwide distribution, however, rarely recorded in the Neotropics; only one

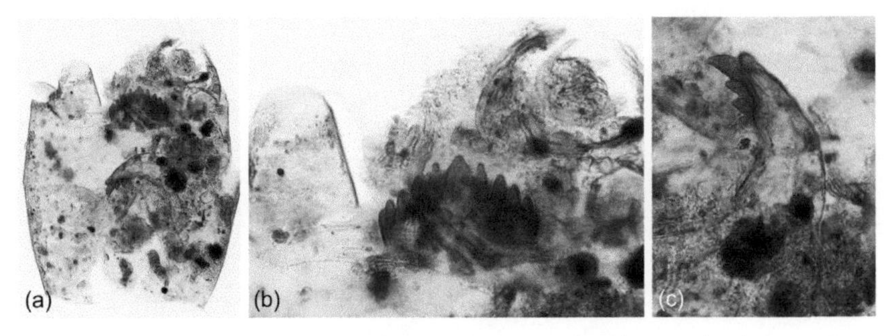

Figure 4.32 *Paratanytarsus penicillatus*-type. a—head capsule, b—mentum, c—mandible.

species has been recorded from Central America (Spies et al. 2009). Larvae live in a variety of aquatic habitats, including brackish water. Some species, such as *Paratanytarsus grimmii*, are notorious pests in water-distribution systems. The recorded morphotype was relatively common in lake sediments across Central America.

RHEOTANYTARSUS THIENEMANN & BAUSE

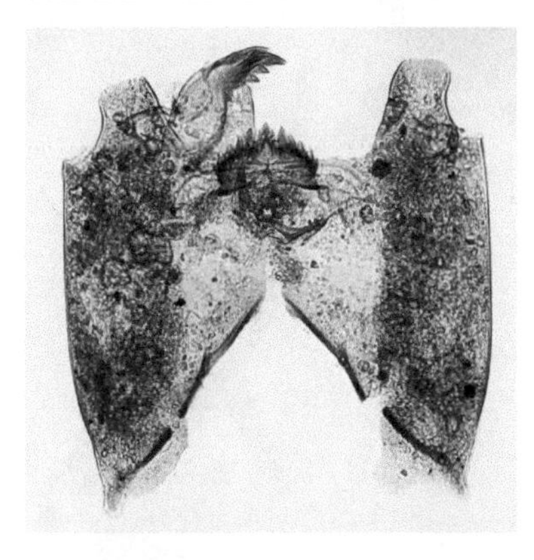

Diagnosis. Mentum with rounded or laterally notched median tooth and 5 pairs of regularly decreasing lateral teeth (Figure 4.33a). Ventromental plates of characteristic shape, strongly curved in outer half and coarsely striated (Figure 4.33c). Post-occipital plate weakly developed. Premandible bifid. Mandible with 2–3 inner teeth, dorsal tooth brown (Figure 4.33a). Antenna 5-segmented, on a short (about as long as broad) pedestal with triangular apical opening, spur absent; Lauterborn organs on short pedicels, not extending beyond antennal apex, opposite on apex of 2nd segment.

Remarks. Rheotanytarsus superficially resembles *Paratanytarsus*, but it can be distinguished by the unique shape of the ventromental plates that are markedly curved and strongly striated. The only type recorded has a laterally notched median tooth and 2 inner teeth on the mandible.

Ecology and Distribution. This is a species-rich genus with worldwide distribution. Larvae are rheobiontic, associated with streams and rivers, rarely can be found in the littoral area of nutrient-poor lakes. Larvae build cases of unique shapes and attach them to various surfaces (Figure 4.33b), including plants, rocks, boats, and aquatic animals. Due to the rheobiontic larvae, *Rheotanytarsus* remains are generally not common in lake sediments and were also rare in Central American lakes.

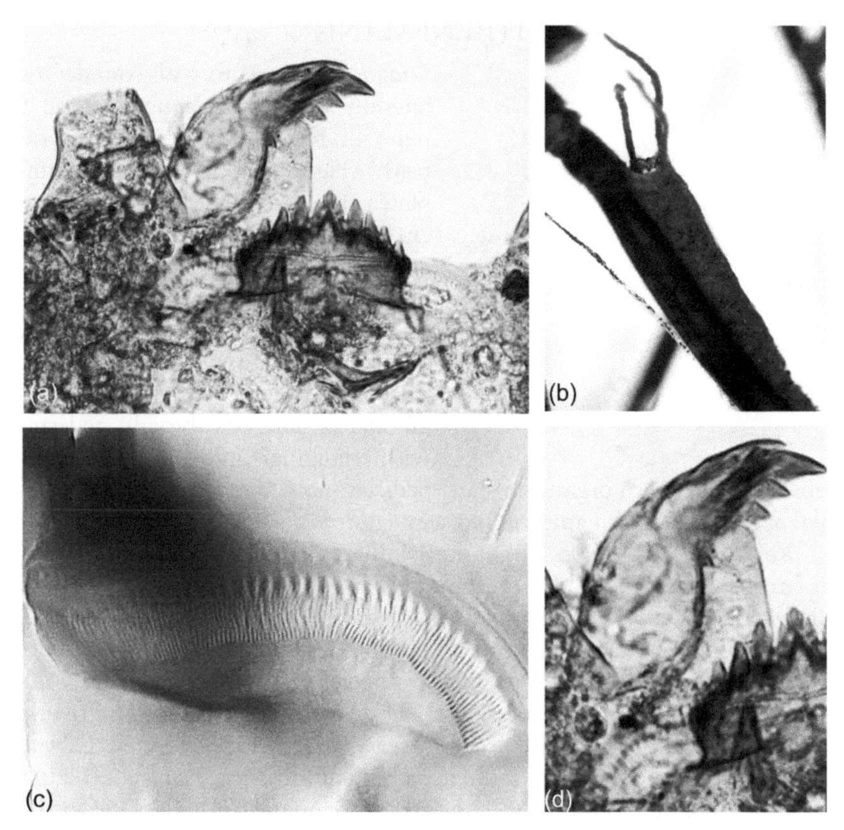

Figure 4.33 *Rheotanytarsus.* a—mentum and mandible, b—larval case (attached to a leg of Odonata), c—detail of ventromental plate, d—mandible modified after Cranston 2010), d—mandible.

STEMPELLINA THIENEMANN & BAUSE

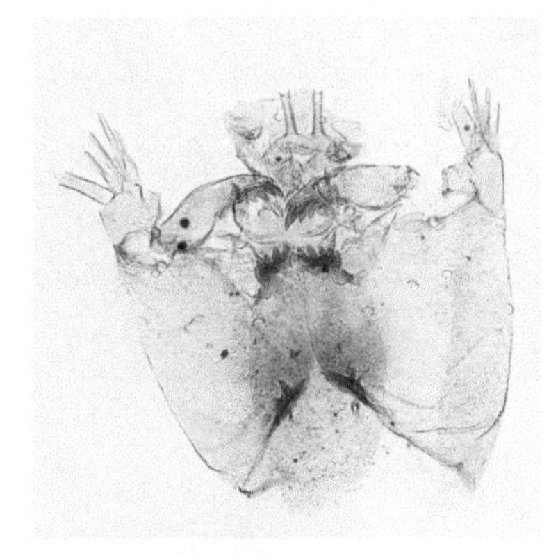

Diagnosis. Mentum with pale rounded or laterally slightly notched median tooth and 6 pairs of darker lateral teeth, regularly decreasing in size (Figure 4.34a,b). Ventromental plates fan-shaped, widely separated medially (Figure 4.34b). Postoccipital plate weakly developed (Figure 4.34a,c). Premandible with 5 teeth. Mandible with brown apical and 2 pointed inner teeth, dorsal tooth yellow. Antenna 5-segmented on prominent pedestal bearing a massive palmate process (Figure 4.34d); Lauterborn organs large, placed on short pedestals.

Remarks. Head capsules of *Stempellina* can be easily separated from other Tanytarsini by the large, palmate expansion on the antennal pedestal.

The recorded head capsules had about 12 prominent knob-like spines on the dorsal surface of the head (Figure 4.34c), which can be classified as *Stempellina johannseni*-type, following Cranston (2010). In addition, the frontoclypeal setae arise from large, elongate pedicels (Figure 4.34d), resembling those of *Stempellina* cf. *subglabripennis* (Brundin) (see Epler 2014 for comparison); however, the recorded morphotype has more dorsal spines, in contrast to *Stempellina* cf. *subglabripennis*, which has only 2.

Ecology and Distribution. A genus with worldwide distribution, rarely recorded in Central America, however, known from Mexico and Guatemala (Spies et al. 2009). Larvae build transportable cases and are found in lentic and lotic habitats. Relatively frequent in Central America as subfossil remains.

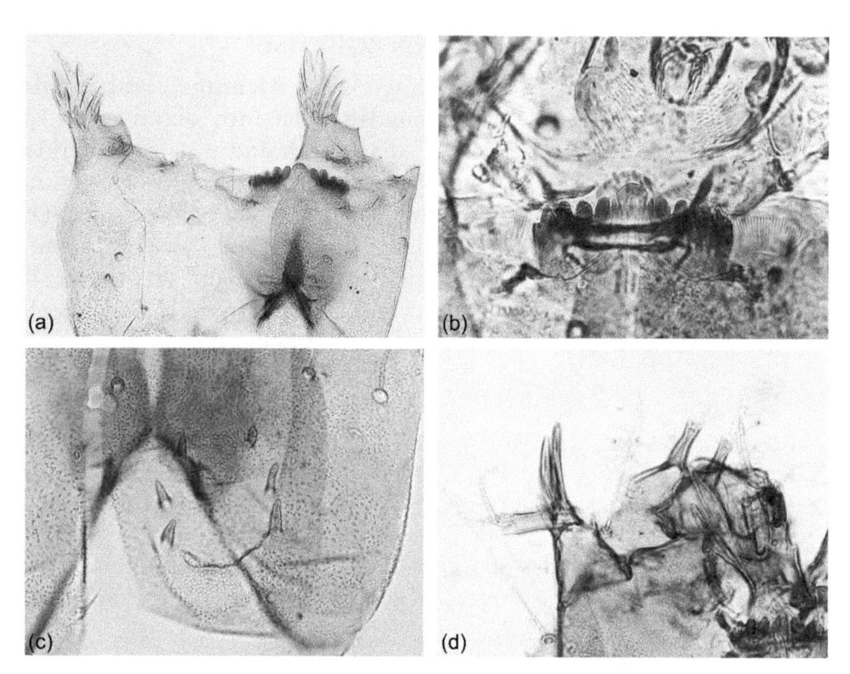

Figure 4.34 *Stempellina*. a—head capsule showing mentum, pedestals with robust palmate processes, and weakly developed post-occipital plate, b—mentum, c—detail of dorsal surface of head with knob-like spines, d—detail of labrum showing elongate pedicels of frontoclypeal setae.

TANYTARSUS VAN DER WULP

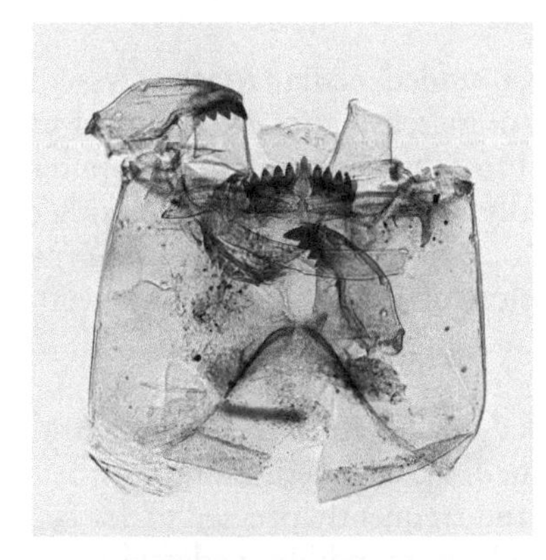

Diagnosis. Mentum with a single median tooth that can be rounded to laterally deeply notched, 5 pairs of lateral teeth present (Figure 4.35b,e,h). Ventromental plates close medially, wide and short, finely striated. Post-occipital plate present, well developed (Figure 4.35a,i,k). Premandible with 3–5 teeth. Mandible with 2–3 inner teeth (Figure 4.35c,j,l), 1–2 dorsal teeth and in some species/morphotypes 1–2. Antenna 5-segmented, on high pedestal usually without spur, however, in some species, a spur may be present (Figure 4.35g); Lauterborn organs on long stems, placed opposite on apex of 2nd segment, occasionally may be annulated.

Remarks. *Tanytarsus* most resembles *Micropsectra* and *Paratanytarsus*, however, it can be distinguished by the premandible with 3 or more apical teeth, as opposed to 2 in *Paratanytarsus* and *Micropsectra.* In the absence of premandible, *Tanytarsus* can be distinguished from *Paratanytarsus* by the length of antennal pedestals, which is short in the latter. Moreover, *Paratanytarsus* is characterized by deeply incised post-occipital arch with weakly developed or missing plate, which also can be used to separate the genus from *Tanytarsus.* *Tanytarsus* can be distinguished from *Micropsectra* by the shape of the spur on the antennal pedestal, usually short and sharp in the latter, while of different shape, if present, in *Tanytarsus.*

KEY TO MORPHOTYPES

1 Mentum with a characteristic shape, rounded median tooth recessed, 5 lateral teeth with a specific pattern: increasing in size from 1st to 3rd (which is the highest), from 3rd to 5th decreasing in size; outer lateral teeth compressed and partially superimposed on each other .... *Tanytarsus norvegicus*-type

1' Mentum without such a combination of characters, lateral teeth decreasing in size gradually 2

2 Mentum with median tooth rounded with a chitinized arch mark, 1st lateral teeth almost as high as median, rounded, 2nd to 5th lateral teeth pointy, 2nd lateral short and frequently pressed to 1st lateral; 2 inner teeth and a dorsal tooth on mandible, pedestal short and with a short thorn bearing spines........... *Tanytarsus* type Yojoa

2' Mentum with median tooth without a chitinized arch 3

3 Antennal pedestal with conspicuous apically rounded thorn; mandible with 2 inner teeth; post-occipital plate weakly developed .... *Tanytarsus ortoni*-type

3' Antennal pedestal with inconspicuous thorn, mandible with 2–3 inner teeth 4

4 Mandible with 2 inner teeth; broad and pale post-occipital margin, plate with slightly concave to straight margin ..... *Tanytarsus* type A

4' Mandible with 3 inner teeth; post-occipital plate dark and strongly concave .... *Tanytarsus* type B

ADDITIONAL REMARKS TO MORPHOTYPES

Five morphotypes were distinguished. *Tanytarsus norvegicus*-type (sensu Lin et al. 2018; Figure 4.35a,b). Mentum with a characteristic shape, rounded median tooth recessed, paler than 5 lateral teeth, their size increasing from 1st to 3rd, which is the highest; from 3rd lateral tooth decreasing in size; outer lateral teeth compressed and partially superimposed on each other. Mandible with a strong apical tooth and a large surficial tooth partially covering inner teeth (their number is not clear), dorsal tooth absent. Post-occipital plate present, antennal pedestal without spur. This morphotype shows features that are shared with *Tanytarsus* and the former *Corynocera*, which was recently moved to *Tanytarsus norvegicus* group. Lin et al. (2018) hypothesize that the latter group originated in Laurasia, which would explain its recent distribution in/near the Arctic and on the Qinghai–Tibet Plateau of the northern hemisphere. Interestingly, the Central American records

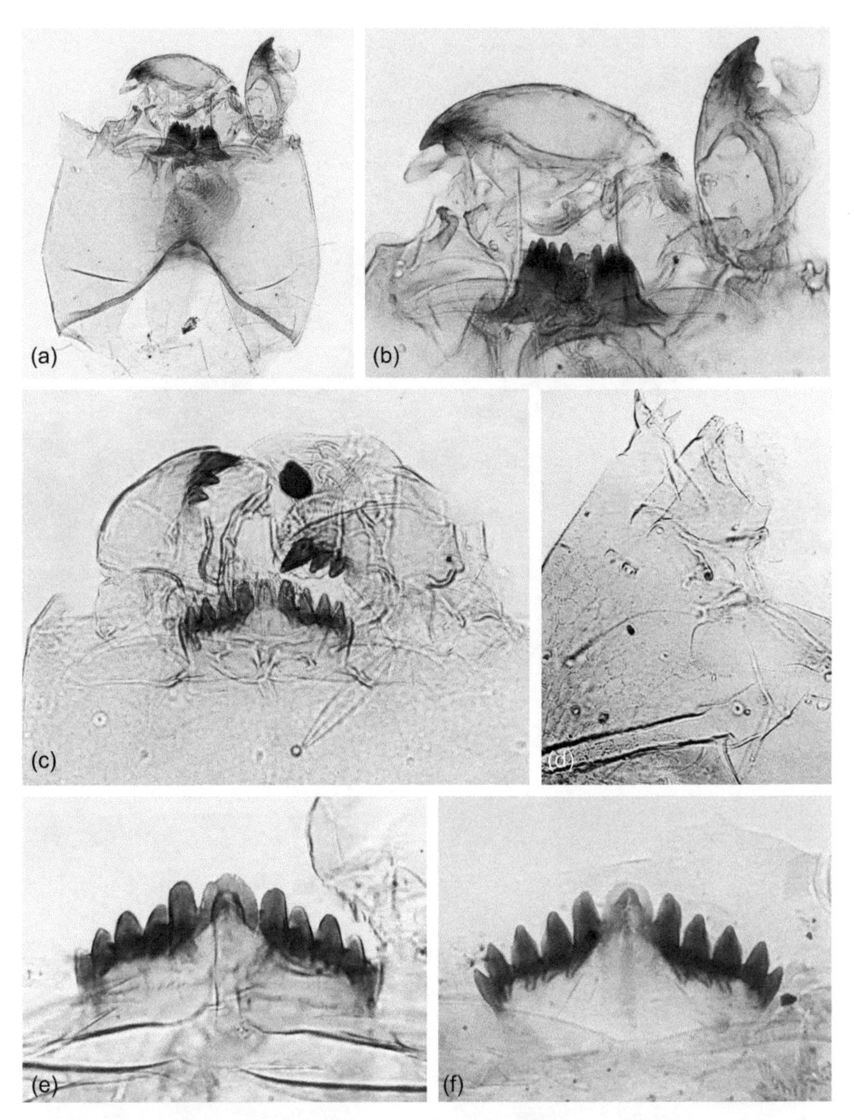

Figure 4.35 *Tanytarsus. Tanytarsus norvegicus*-type: a—head capsule, b—mentum and mandible; *Tanytarsus* type Yojoa: c—detail of head with mentum and mandible, d—detail of head capsule showing reticulation and thorn bearing spines situated on pedestals; e, f—variations of menta; *Tanytarsus ortoni*-type: g—detail of the head capsule with mentum, mandible (and antennal pedestal with spur), h—mentum and mandible; *Tanytarsus* type A: i—head capsule, j—mandible with 2 inner teeth; *Tanytarsus* type B: k—head capsule, l—mandible with 3 inner teeth.

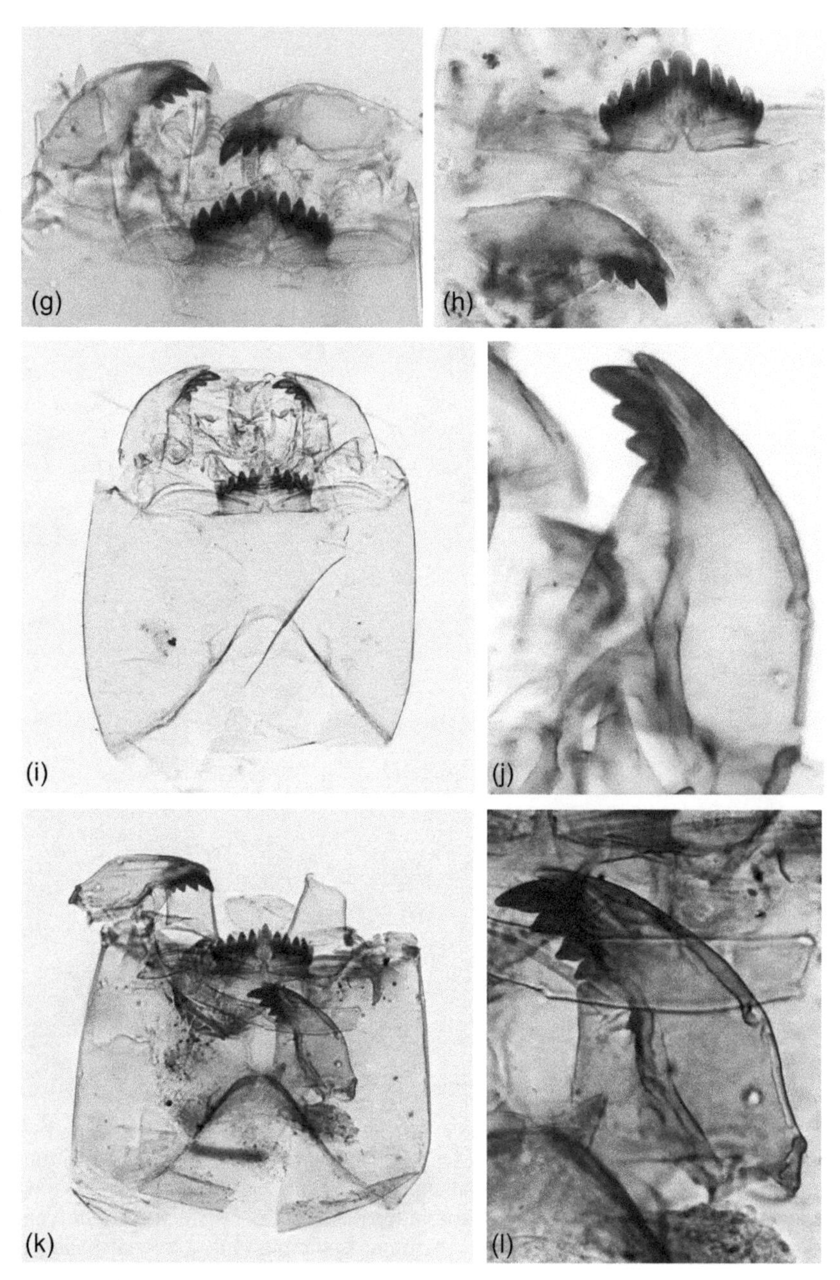

Figure 4.35 *Continued.*

contradict it, since the remains were found in warm lowland lakes in Belize, Guatemala, and Mexico. A similar taxon is known from Florida: Epler (2014) presents an unusual larva showing deformed mandible and mentum called *Tanytarsus* sp. beta; the adults of this type represent an undescribed species.

Tanytarsus type Yojoa (Figure 4.35c–f). It can be distinguished by a characteristic broad and rounded median tooth with a chitinized arch mark, 1st lateral almost as high as median tooth, rounded, 2nd to 5th lateral teeth pointed, 2nd lateral short and frequently appressed to 1st lateral; 2 inner teeth and a dorsal tooth on mandible, pedestal short and with a thorn bearing spines. Head capsules of some specimens with distinct reticulation. Post-occipital plate reduced or missing.

Tanytarsus ortoni-type (sensu Lin et al. 2018; Figure 4.35g,h). Premandible with 3 teeth. Mandible with 2 inner teeth. Antenna pedestal with conspicuous apically rounded thorn variable in size. Post-occipital plate present but weakly developed. This type corresponds to the former genus *Caladomyia* that has been assigned to *Tanytarsus ortoni* group (Lin et al. 2018).

Tanytarsus type A (Figure 4.35i,j). Head pale, mentum with laterally notched median tooth paler than lateral teeth, broad pale occipital margin, plate with slightly concave end. Mandible with 2 inner teeth. Antennal pedestals without spur.

Tanytarsus type B (Figure 4.35k,l). Head capsule pale, dark and strongly concave post-occipital plate. Mandible with 3 inner teeth.

Ecology and Distribution. Tanytarsus is one of the most widely distributed, species-rich, and ecologically diverse chironomid genera, however only a few named species have been recorded from Central America (Spies et al. 2009), with many more species expected to be discovered. Epler (2017), for example, recorded eight undescribed species from Costa Rica. Larvae occur virtually in all types of fresh to saline waters, including some marine species from the western Pacific. Common in Central American subfossil material (Figure 4.35a,b).

TANYTARSINI TYPE A

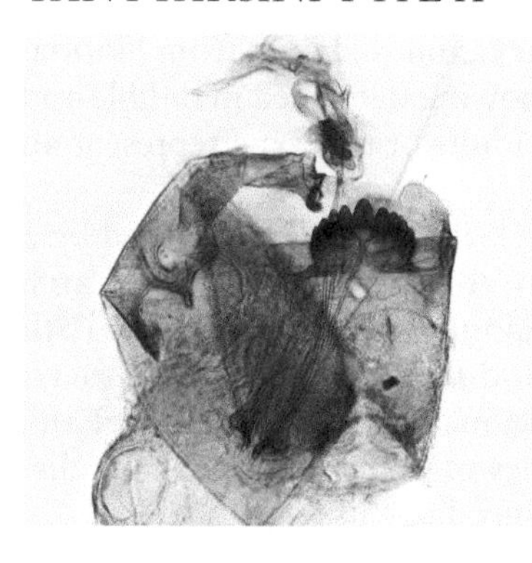

Diagnosis. Head capsule strongly sclerotized, brown. Mentum with rounded median tooth and 4 pairs of lateral teeth; median tooth longer and broader than 1st lateral, central triplet of teeth appear to be on a different focal plane compared to outer lateral teeth; all median teeth uniformly dark (Figure 4.36a). Ventromental plates close medially, each plate slightly shorter than width of mentum. Post-occipital arch is deeply incised, plate present but weakly developed (Figure 4.36b). Premandible with 3 teeth. Mandible and antenna unknown.

Remarks. The generic placement of this morphotype is uncertain, but the characteristic mentum with rounded median tooth and only 4 lateral teeth resemble that of *Neozavrelia*, which has not been recorded in Central America so far.

Ecology and Distribution. Rare and only found in a high elevation lake at ~2000 m a.s.l.

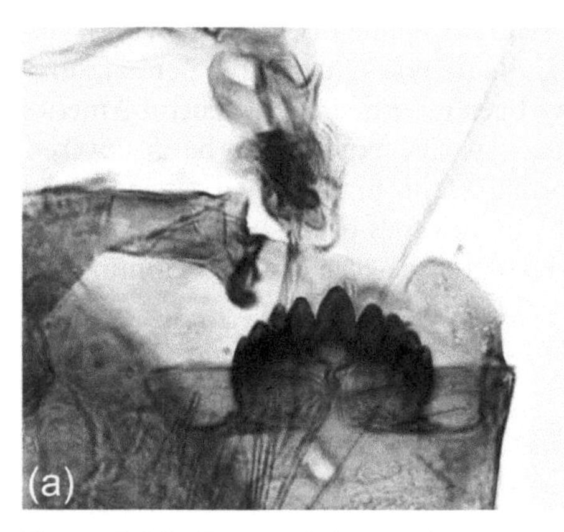

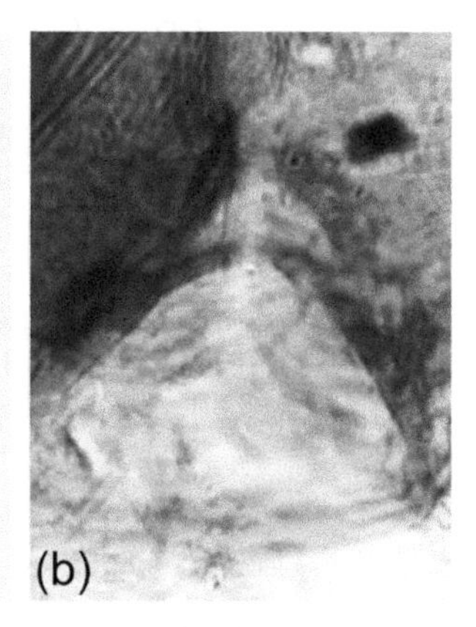

Figure 4.36 Tanytarsini type A. a—mentum, ventromental plates, and premandible, b—post-occipital arch.

Larvae of Orthocladiinae show great morphological variability but can be separated from Chironominae by the eye-spots, which are often contiguous or, when separate, with a dorsal eye-spot posterior to a ventral eye-spot (Andersen et al. 2013). Another distinctive trait in the larval stage is the absence of striated ventromental plates that are usually indistinct, although some *Nanocladius* may bear horizontal striae. There is no accepted division of Orthocladiinae into tribes. Figure 5.1 shows Orthocladiinae larval head capsule with characteristic subfossil features.

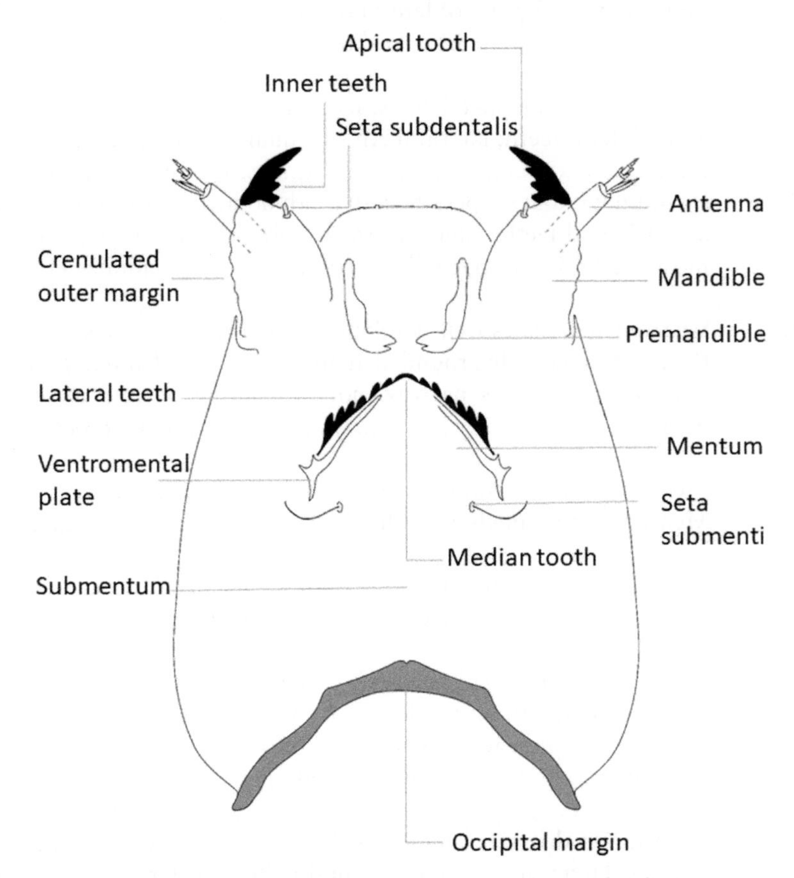

Figure 5.1 Orthocladiinae larval head capsule showing subfossil characters.

DOI: 10.1201/9780429021572-5

KEY TO GENERA OF ORTHOCLADIINAE*

1 Mentum with single median tooth...2
1' Mentum with 2 or more median teeth..6

2(1) Mentum with 4 pairs of lateral teeth............ *Eukiefferiella* in part
2' Mentum with 5–6 pairs of lateral teeth...3

3(2) Mentum with 5 pairs of lateral teeth..4
3' Mentum with 6 pairs of lateral teeth..5

4(3) Mentum with distinct stripes, median tooth about 4× as broad as 1st lateral teeth, lateral teeth subequal in size........................ ....*Eukiefferiella* in part
4' Mentum without stripes, median tooth domed, up to 3× as broad as 1st lateral teeth; lateral teeth sharply decrease in size, outermost 2 lateral teeth partially fused............ Orthocladiinae type A

5(3) Ventromental plates well-developed, broad and extending beyond the mentum laterally, rounded at apex.................*Parakiefferiella*
5' Ventromental plates narrow, not protruding beyond mentum laterally..*Cricotopus/Orthocladius*

6(1) Mentum with 2 median teeth...7
6' Mentum with 3 median teeth...17

7(6) Mentum with 4 pairs of lateral teeth......................................8
7' Mentum with 5 or more pairs of lateral teeth.........................10

8(7) Mentum of prolonged triangular shape, median teeth of mentum very high, about the length of 4 lateral teeth; lateral teeth separated from each other by a narrow notch, ventromental plates long and narrow, extending well beneath mentum ....*Synorthocladius*
8' Mentum of different shape, median teeth only slightly higher than 1st lateral tooth, ventromental plates shorter...................9

9(8) Mentum with 2 elongate median teeth, small central tooth may be present between median teeth, outermost lateral tooth may be notched apically...*Brillia***
9' Mentum with medial teeth broad and domed, ventromental plates well-developed, extending beyond lateral margin of mentum...................*Bryophaenocladius/Gymnometriocnemus*

10(7) Ventromental plates well-developed, extending beyond lateral margin of mentum..11
10' Ventromental plates vestigial, if present, narrow and not extending beyond mentum laterally..15

11(10) Ventromental plates bulbous at apex, mentum rather horizontal ....*Heterotrissocladius*
11' Ventromental plates triangular, foot-shaped or truncated 12

12(11) Ventromental plates very long, extending below the mentum by at least half the lengths of the mentum, 5–6 lateral teeth on mentum..*Nanocladius*
12' Ventromental plates much shorter 13

13(12) Ventromental plates foot-shaped, 3rd lateral mental tooth shorter than 4th .... *Parametriocnemus/Parachaetocladius*
13' Ventromental plates triangular or truncated.............................14

14(13) Ventromental plates triangular at apex, beard, if present, weak, consists of few short setae.................................. *Psectrocladius*
14' Ventromental plates truncate at apex, with dense long beard, a small accessory tooth can be present between median and 1st lateral tooth..*Rheocricotopus*

15(10) Mentum with bulbous, darkly pigmented tubercle situated on each side of the base.....................*Limnophyes/Paralimnophyes*
15' Mentum without such tubercles laterally at the base 16

16(15) Mentum with median teeth higher than 1st lateral tooth, and with distinct stripes.. *Eukiefferiella* in part
16' Mentum with median teeth shorter than 1st lateral tooth .... *Metriocnemus*

17(16) Antenna 4-segmented, as long or longer than head capsule ....*Corynoneura*
17' Antenna 5-segmented, half to $^{1}/_{2}$ of the length of head.............. ....*Thienemanniella*

* Modified after Bitušík (2000), Brooks et al. (2007) and Andersen et al. (2013).

** In fact, *Brillia* has 5 lateral teeth, but the 4th and 5th are small and fused, appearing as one notched tooth.

BRILLIA KIEFFER

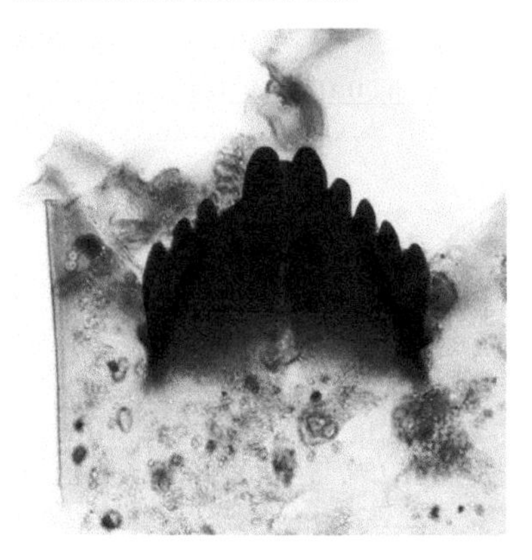

Diagnosis. Mentum with a median complex consisting of a minute central tooth between bases of pair of large median teeth (Figure 5.2); 5 pairs of lateral teeth, inner 3 lateral teeth of the same size and shape, 4th and 5th lateral teeth smaller and appressed to each other, looking like a notched tooth. Seta submenti posterior to the mentum. Mandible with short apical tooth and 4 inner teeth (4th one blunt and hardly distinguishable). Antenna 4-segmented; segment 2 divided into 2 sections. Premandible weakly bifid at apex.

Remarks. There are some taxonomic uncertainties in the *Brillia*-complex, thus the present head capsule can belong to one of the genera closely related to *Brillia* with unknown immature stages, e.g., *Irisobrillia*, which has been recorded in the Neotropical region (Andersen and

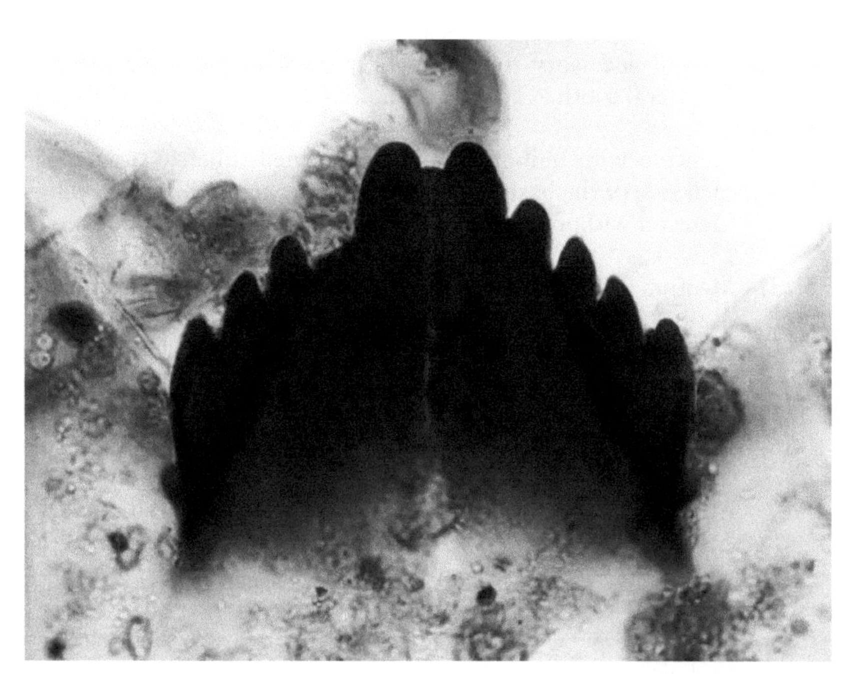

Figure 5.2 *Brillia*. Mentum.

Mendes 2004) and in Central America (Spies and Reiss 1996; Watson and Heyn 1992; Epler 2017).

Ecology and Distribution. Larval *Brillia* live in a broad variety of environments from littoral of lakes and hygropetric habitats to running waters. Some species mine in immersed wood, others live in dead leaves. *Brillia* head capsules were rare in Central America, restricted to high elevations.

BRYOPHAENOCLADIUS GOETGHEBUER / *GYMNOMETRIOCNEMUS* THIENEMANN

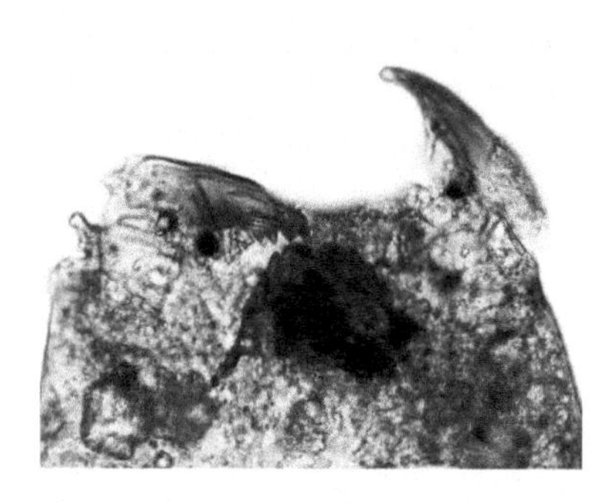

Diagnosis. Mentum with 2 (occasionally 1) broadly rounded median teeth and 4 pairs of lateral ones (Figure 5.3a,b). Ventromental plate heavily sclerotized but not extending beyond the lateral margin of flattened mentum; beard absent. Antenna 5-segmented. Premandible with 3 teeth. Mandible with short apical tooth and 3 inner teeth (Figure 5.3a).

Remarks. Recent larvae of *Gymnometriocnemus* can only be distinguished from *Bryophaenocladius* by the faintly divided posterior parapods (*Bryophaenocladius* has parapods undivided). Regarding subfossil material, if the body is missing, the remains of these two genera cannot be distinguished.

Ecology and Distribution. *Bryophaenocladius* is species-rich and worldwide in distribution. The biology of the genera *Bryophaenocladius*/*Gymnometriocnemus* is fragmentary, the known species are mainly terrestrial but several species are aquatic or semiaquatic, inhabiting lake shorelines, seeps, and springs (Epler 2001). Rare in the Central American samples, which may be a consequence of the possible terrestrial/semiaquatic habit of the larvae.

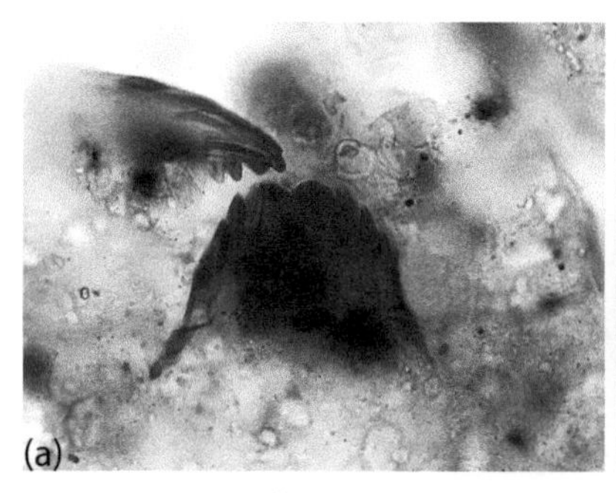

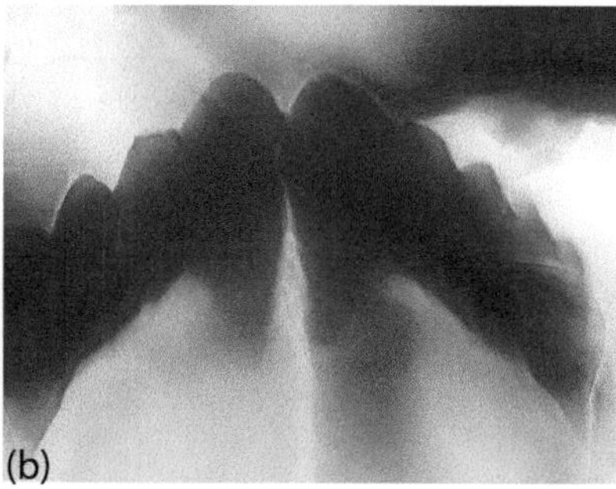

Figure 5.3 *Bryophaenocladius/Gymnometriocnemus*. a—mentum and mandible, b—mentum with well-developed ventromental plate (modified after Cranston 2010).

CORYNONEURA WINNERTZ

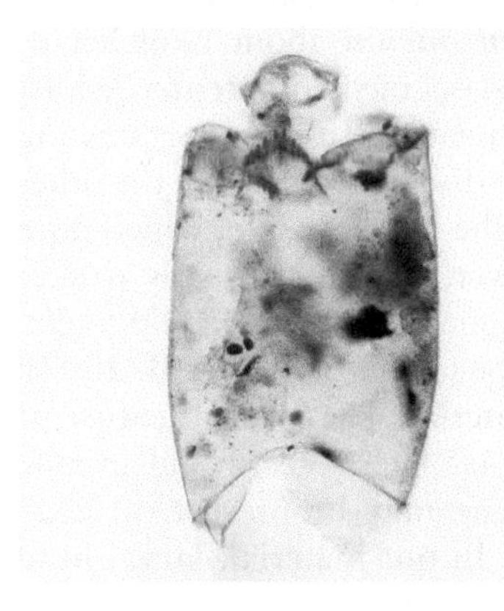

Diagnosis. Head capsule tiny, narrow and prolonged (Figure 5.4a). Mentum strongly triangular in shape, with 2–3 median teeth, central tooth, if present of various sizes from minute to subequal to outer median teeth (Figure 5.4b–d); 5 pairs of lateral teeth. Premandible broad with many short teeth at the apex. Mandible with short apical tooth, usually subequal or shorter than 1st inner teeth, 4 inner teeth present (Figure 5.4b). Antenna is long or longer than head capsule, 4-segmented, 1st segment elongate, segments 2 and 3 often darkened.

Remarks. *Corynoneura* resembles *Thienemanniella* and *Onconeura*. The distinctive feature is the antennal length: in *Corynoneura* the

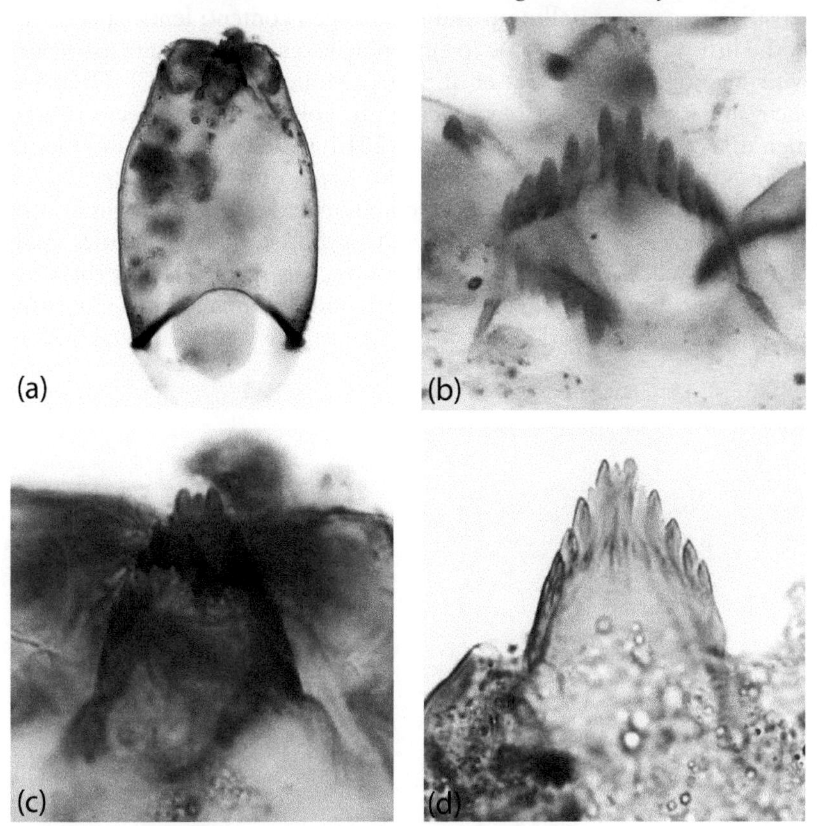

Figure 5.4 *Corynoneura*. a—head capsule, b, c, d—variations of menta.

antenna is as long or longer than head, while in *Thienemanniella* it is about ½ to ¾ lengths of head, and in *Onconeura* about ⅓ of length of head. In addition, *Corynoneura* has a 4-segmented antenna, whilst *Thienemanniella* and *Onconeura* 5-segmented. In case of missing antenna, some *Corynoneura* species can be distinguished from the other two genera by presence of reticulation of the head capsule, which does not occur in the other two genera. Moreover, heads are usually pale in *Corynoneura*, while generally strongly pigmented in *Thienemanniella*. One morphotype, *Corynoneura lobata*-type (sensu Brooks et al. 2007), was recorded in the Central American material. The typical feature of this morphotype is the strongly reduced 1st lateral teeth and minute central median tooth. Wrinkled sculpturing may be present on head capsule, but reticulation is inconspicuous. In our material, larval head capsules did not possess sculpturing or reticulation.

Ecology and Distribution. *Corynoneura* is a species-rich genus with worldwide distribution. Larvae live in various types of waterbodies, in lakes, streams, rivers, from lowland to high mountains and arctic waters. In the Palaearctic, lotic species are considered oligostenothermal, and sensitive to organic pollution and low oxygen content; lentic species live in the littoral of oligotrophic to eutrophic lakes and are often associated with macrophytes (Brooks et al. 2007; Moller Pillot 2013). Owing to their small size, larval *Corynoneura* are polyvoltine and seem to have high dispersal capacity (Moller Pillot 2013) making them ready colonizers of temporary waters. Together with *Cricotopus* and *Psectrocladius*, larval *Corynoneura* were the most common orthoclads found in Central American lakes. Due to its small size, *Corynoneura* head capsules might be overlooked in the subfossil (and also recent) material, thus their real distribution and taxonomic richness are underestimated both in Central America as well as worldwide.

CRICOTOPUS VAN DER WULP / *ORTHOCLADIUS* VAN DER WULP

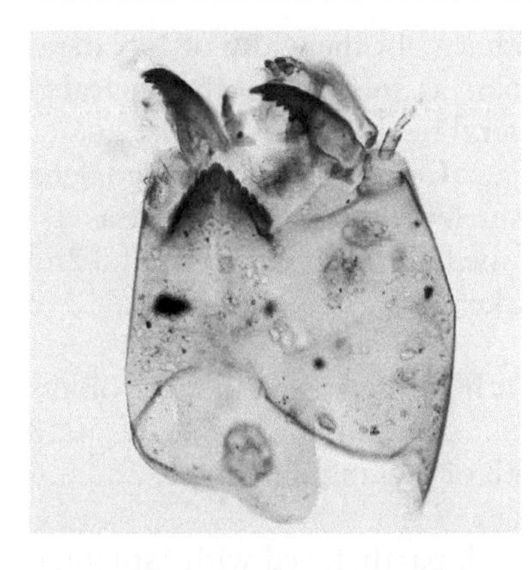

Diagnosis. Be aware that larvae of *Cricotopus* are highly similar and often indistinguishable from *Orthocladius* even in recent larvae and it is even more true about the subfossil remains.

Mentum with single median tooth of different size, and usually 6 pairs of lateral teeth (Figure 5.5a,d,g), however, a few species of *Cricotopus* have 5 or 7 pairs of lateral teeth, while some *Orthocladius* may have 7–9 pairs. Mandible with a relatively short apical tooth and 3 inner teeth (Figure 5.5b,d). Premandible with 1–2 apical teeth. Antenna usually 5-segmented (seldom 4-segmented), some species have reduced antenna.

Remarks. Both *Cricotopus* and *Orthocladius* have a mentum with single median tooth and (usually) 6 lateral teeth, mandible with short apical tooth and 3 inner teeth. While in recent larvae the tuft of setae on the body segments of most of the species can distinguish *Cricotopus* from *Orthocladius*, due to the lack of a body, this feature cannot be used for subfossil remains. However, the following characters, in general, seem to be more typical for *Cricotopus* than *Orthocladius*: outer margin of mandible strongly crenulated, 2nd lateral tooth is smaller than 1st and/or reduced and fused to 1st lateral tooth, and outer four lateral teeth forming a distinct group.

KEY TO MORPHOTYPES

1 Mentum with broad pale median tooth (ca. 4× the width of 1st lateral tooth); 1st lateral tooth the same color as median tooth, separated from it by a narrow notch; outer 5 lateral teeth darker........................ .... *Cricotopus* type Magdalena

1' Mentum with median tooth narrower (at most 3× as broad as 1st lateral tooth); mentum unicolored, if median tooth paler, 1st and 2nd teeth as well, outer 4 lateral teeth darker..2

2 Mentum with lateral teeth rounded with pointed apex, of more or less triangular shape...3

2' Mentum with at least 2–6 lateral teeth of rectangular shape............6

3 Mentum with 2nd lateral tooth reduced, partly fused with 1st lateral tooth..4

3' Mentum with 2nd lateral tooth not reduced (although may be slightly smaller than adjacent teeth) and never fused with 1st lateral tooth.......5

4 Mentum strongly pigmented, uniformly dark; median tooth slightly higher and about 2× as broad as 1st lateral tooth.............................. .... *Cricotopus* type I

4' Mentum with conspicuous color pattern: median and first 2 lateral teeth pale, 4 outer lateral teeth darker; median mental tooth rounded, considerably higher than 1st lateral tooth; outer margin of mandible strongly crenulated.. *Cricotopus* type Atitlan

5 Mentum light brown; median tooth slightly higher and wider than 1st lateral; 1st lateral tooth prolonged, all teeth slender, higher than wide .... *Cricotopus* type A

5' Mentum strongly sclerotized, dark brown; median tooth domed, considerably higher than 1st lateral, which is circular in shape .... *Cricotopus* type B

6 Head capsule yellow; median tooth subequal to 1st lateral tooth; 1st lateral teeth circular, broader in the middle than at the base .... *Cricotopus* type C

6' Head capsule dark brown; median mental tooth with straight anterior margin, about 2× as broad and about as high as 1st lateral; 1st lateral larger than outer 5 lateral teeth of rectangular shape; mental extensions protracted posteriorly beyond seta submenti..................... .... *Orthocladius* (*Euorthocladius*) sp.

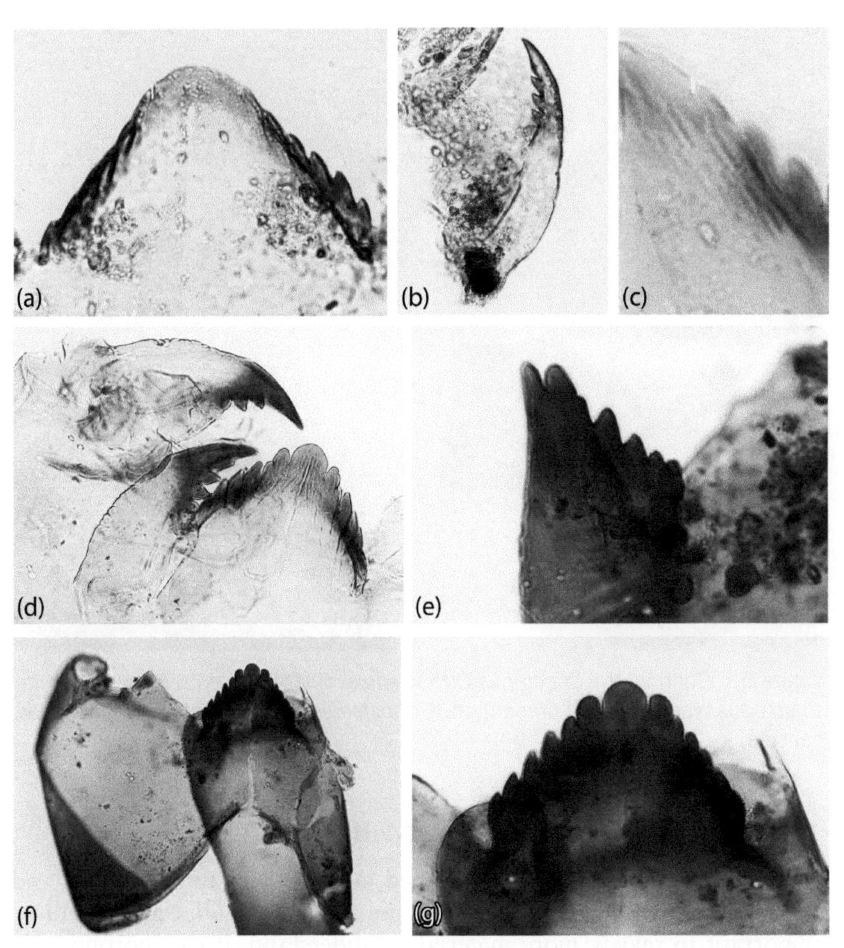

Figure 5.5 *Cricotopus/Orthocladius*. *Cricotopus* type Magdalena: a—mentum, b—mandible, c—detail of mentum showing the notch between median and 1st lateral tooth; *Cricotopus* type Atitlan: d—mentum and mandibles; *Cricotopus* type I: e—half of mentum; *Cricotopus* type B: f—head capsule, g—mentum.

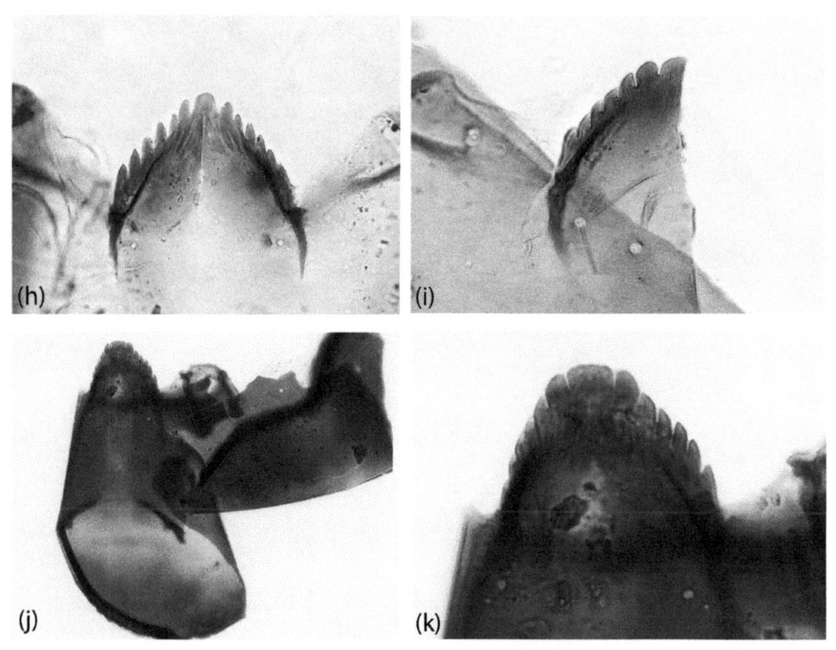

Figure 5.5 Continued *Cricotopus/Orthocladius. Cricotopus* type A: h—mentum; *Cricotopus* type C: i—half of mentum; *Orthocladius* (*Euorthocladius*) sp.: j—head capsule, k—mentum.

ADDITIONAL REMARKS TO MORPHOTYPES

Seven morphotypes were distinguished, six belonging to *Cricotopus* and one to *Orthocladius*. Some of the types are quite similar and it will be important to review more material to understand, if the morphotypes are valid or just variations within a gradient of morphological variability.

Cricotopus type Magdalena (Figure 5.5a–c). Broad pale median tooth, 6 lateral teeth, 1st lateral as pale as median tooth, 5 outer lateral darker; 1st lateral tooth separated from the median one with a narrow notch. Mandible with long slender apical tooth and 3 inner teeth. The morphotype shows some similarity with *Cricotopus cylindraceus*-type and *Cricotopus* type P sensu Brooks et al. (2007).

Cricotopus type Atitlan (Figure 5.5d). Median mental tooth rounded, considerably higher than 1st lateral tooth; 2nd lateral tooth small and partly fused with the 1st lateral tooth; median and first 2 lateral teeth pale, outer 4 lateral teeth darker. Premandible weakly bifid at apex. Mandible with 3 inner teeth, outer margin strongly crenulated. Similar to *Cricotopus sylvestris* group sensu Andersen et al. (2013) but median tooth considerably broader.

Cricotopus type I (Figure 5.5e). Head capsule brown, mentum strongly pigmented, uniformly colored. Mentum with median tooth slightly higher and about 2× as broad as 1st lateral tooth; 2nd lateral tooth minute and partly fused to 1st lateral. Similar to *Cricotopus intersectus*-type sensu Brooks et al. (2007), but the whole head capsule is dark.

Cricotopus type A (Figure 5.5f). Mentum light brown; median tooth slightly higher and wider than 1st lateral tooth; 2nd lateral tooth somewhat smaller than 1st and 3rd lateral. All teeth are slender, higher than wide.

Cricotopus type B (Figure 5.5g,h). Mentum strongly sclerotized, dark brown; median tooth domed, considerably higher than 1st lateral, which is semi-circular in shape, 2nd lateral tooth smaller than 1st and 3rd lateral teeth.

Cricotopus type C (Figure 5.5i). Mentum with median tooth subequal to 1st lateral tooth; 1st lateral semi-circular, broader in the middle than at the base. Outer 5 lateral teeth subequal, rectangular in shape, creating an apically flattened outline.

Orthocladius (*Euorthocladius*) sp. (Figure 5.5j,k). Head capsule strongly pigmented, dark brown; mentum of the same color as head. Median mental tooth with straight anterior margin, about 2× as broad and about as high as 1st lateral; 1st lateral rounded, broader and higher than outer 5 lateral teeth of rectangular shape; mental extensions extended posteriorly beyond seta submenti; occipital margin dark brown.

Ecology and Distribution. Cricotopus is one of the largest Orthocladiinae genera with worldwide distribution (Cranston et al. 1989; Ashe and O'Connor 2009). *Orthocladius* is also widespread, with a few morphologically diverse subgenera and numerous species in the Holarctic region, where they live in all kinds of aquatic habitats, but rarely reported from the Neotropics (Spies et al. 2009). Reported from lotic waters of Costa Rica (Watson and Heyn 1992) and from the Dominican Republic (Silva et al. 2015b). Larvae of both genera occur in all types of waterbodies, including some saline and coastal waters. The group is associated with aquatic macrophytes, algae, and sometimes cyanobacteria. Only a small fragment of the larvae of the Neotropics is known (Andersen et al. 2013). *Cricotopus* was the most frequent and abundant orthoclad in Central American lake sediments, occurring at both low and high elevations.

EUKIEFFERIELLA THIENEMANN

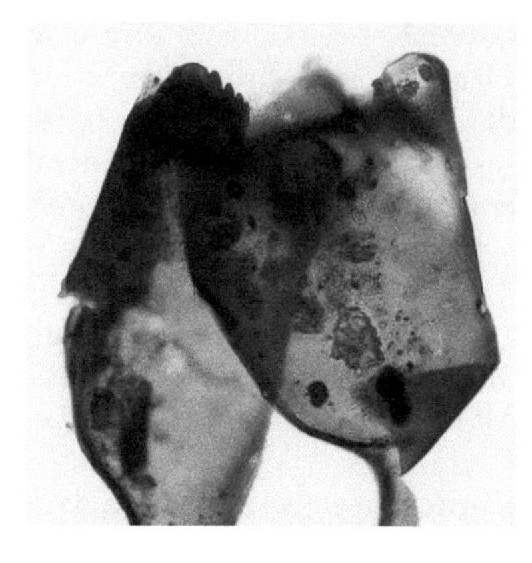

Diagnosis. Head capsule usually dark and strongly pigmented with dark brown to black occipital margin (Figure 5.6a). Mentum with vertical striped (Figure 5.6b), except for *Eukiefferiella devonica*-type (Figure 5.6d), with 1–2 median and usually 5 (4 in *Eukiefferiella devonica*-type) pairs of narrow, pointed lateral teeth. Mandible with short apical tooth and 3–4 inner teeth (Figure 5.6e); mola (inner margin of mandible) with up to 5 short spines. Antenna usually 5-segmented (occasionally 4-segmented). Premandible broad with a single apical tooth.

Remarks. The most characteristic feature distinguishing *Eukiefferiella* from other genera is the striped mentum and spines on the inner margin of mandible.

KEY TO MORPHOTYPES

1 Mentum with single broad median tooth... 2
1' Mentum with bifid median tooth........ *Eukiefferiella claripennis*-type

2 Mentum with 4 lateral teeth .... *Eukiefferiella devonica*-type
2' Mentum with 5 lateral teeth .... *Eukiefferiella gracei*-type

ADDITIONAL REMARKS TO MORPHOTYPES

Three morphotypes were distinguished according to Brooks et al. (2007) and Andersen et al. (2013).

Eukiefferiella claripennis-type (Figure 5.6a,b). It has a mentum distinctly striated with bifid median tooth (separated by narrow notch), and 5 lateral teeth; 1st lateral appressed to median teeth.

Eukiefferiella devonica-type (Figure 5.6c,d). Characteristic by a mentum with single broad median tooth and 4 lateral teeth; median tooth 2–3 times wider than 1st lateral. Mentum uniformly dark, strongly pigmented, striation not visible.

Eukiefferiella gracei-type (after Andersen et al. 2013, Figure 5.6e,f). This morphotype possesses a mentum with vertical stripes, and a single broad median tooth and 5 lateral teeth. It is similar to *Eukiefferiella fittkaui*-type sensu Brooks et al. (2007). However, *Eukiefferiella fittkaui*, which also belongs to the *gracei*-species group, has a Palaearctic

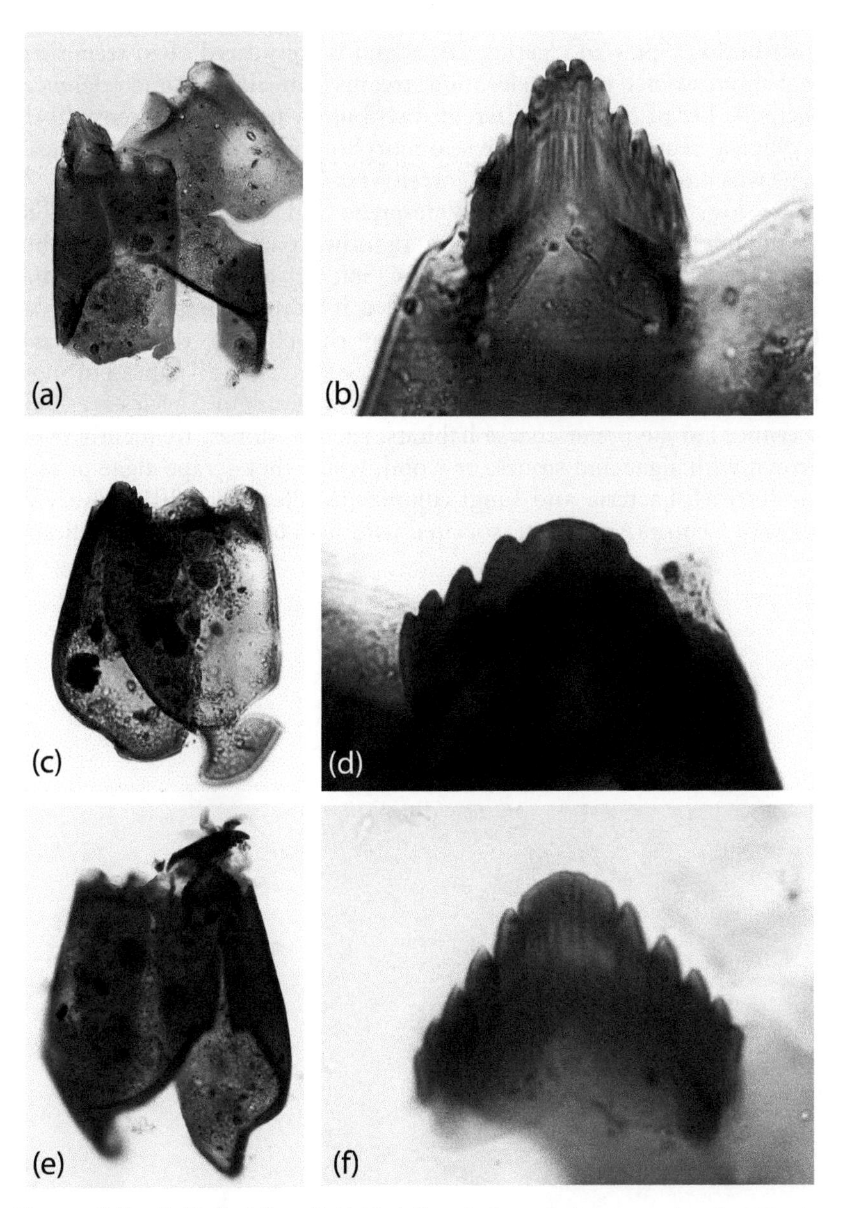

Figure 5.6 *Eukiefferiella. Eukiefferiella claripennis*-type: a—head capsule, b—mentum; *Eukiefferiella devonica*-type: c—head capsule, d—mentum; *Eukiefferiella gracei*-type: e—head capsule, f—mentum.

distribution (Spies and Sæther 2013) and is considered oligo-stenothermal and restricted to high elevation streams (Bitušík 2000 and references therein). Because of the different distribution pattern and most likely ecological requirements of the two morphotypes, the Central American type was named *Eukiefferiella gracei*-type.

Ecology and Distribution. Widespread and species-rich genus with worldwide distribution. Common in other parts of the world, but rarely recorded from the Neotropics (Spies et al. 2009). Watson and Heyn (1992) and Epler (2017) recorded it from rivers and streams of Costa Rica. Larvae of *Eukiefferiella* are rheophilic to rheobiontic and oligostenothermic associated with flowing waters of all types but most frequently with colder montane waters; however, some species are eurythermic. Larvae prefer coarse habitats, such as stones, frequently overgrown with algae and stones, or wood, where they scrape algae or feed on detritus, bacteria and fungi colonies (Moller Pillot 2013). Rare in lakes of Central America, associated with high elevation waterbodies.

HETEROTRISSOCLADIUS SPÄRCK

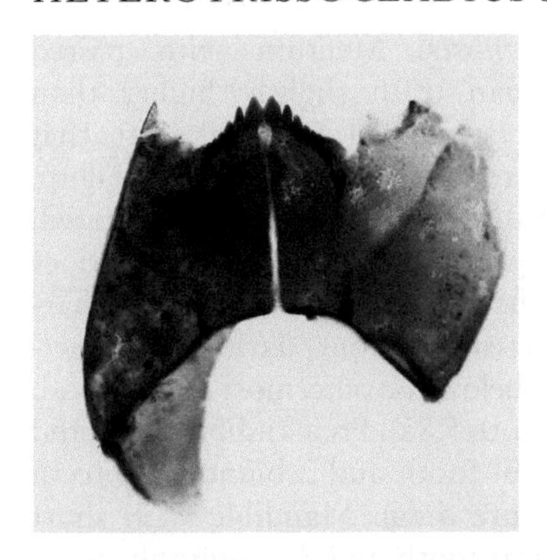

Diagnosis. Mentum usually with 2 median teeth, single in *Heterotrissocladius maeaeri*-type, and 5 pairs of lateral teeth (Figure 5.7a); median teeth may be laterally notched. Ventromental plate distinct, extended beyond margin of flattened mentum. In some morphotypes, such as *Heterotrissocladius marcidus*-type, the submentum is darker than the rest of the head (Figure 5.7b). Premandible with a single tooth. Mandible with a long, slender apical tooth and 3–4 inner teeth. Antenna 7-segmented; 3rd segment much smaller than 4th; segment 7 hair-like, hardly visible.

Remarks. Heterotrissocladius can be distinguished from genera with mentum with 2 median teeth and 5 pairs of lateral teeth by the strong bulbous ventromental plates, extending beyond the lateral margin of the mentum and the lack of beard. In addition, the darker submentum, typical for *Heterotrissocladius marcidus*-type, is also unique. Only one morphotype was recorded, *Heterotrissocladius marcidus*-type.

Ecology and Distribution. In general, larval *Heterotrissocladius* colonize different waterbody types, littoral and profundal of lakes, ponds, streams and rivers. The genus is cold-stenothermal and is associated with oligotrophic lakes with *Heterotrissocladius marcidus*-type being the least thermally restricted morphotype (Sæther 1975). Rare morphotype, associated to high elevation lakes. This is the first record from Central America (Hamerlík and Silva 2018) and the second for the Neotropical region (Turcotte and Harper 1982).

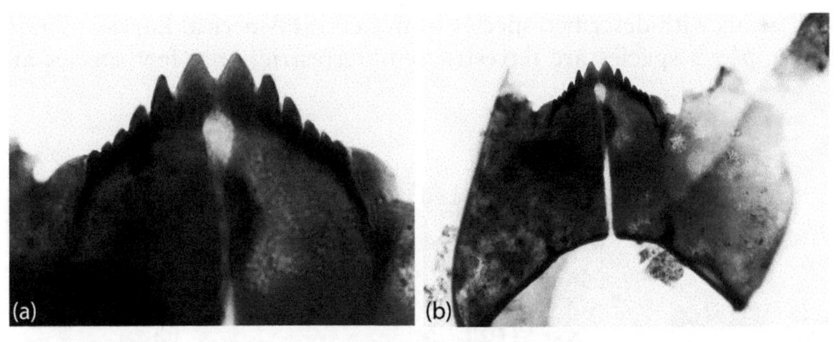

Figure 5.7 *Heterotrissocladius marcidus*-type. a—mentum, b—head capsule.

LIMNOPHYES EATON /*PARALIMNOPHYES* BRUNDIN

Diagnosis. Mentum with paired median teeth slightly higher than the 1st pair of 5 lateral teeth that descrease in size gradually (Figure 5.8a); mental teeth darkly pigmented, frequently with pale vertical stripes in the median region; there is a characteristic bulbous, dark lobe projecting below the outermost lateral tooth (Figure 5.8a). Premandible with bifid apical tooth and 2 blunt inner teeth (Figure 5.8c). Mandible with short apical tooth and 3 (*Limnophyes*) or 4 (*Paralimnophyes*) inner teeth (Figure 5.8b). Antenna 5-segmented, relatively short, about ½ of length of mandible, 4th segment longer than 3rd.

Remarks. The only way to distinguish *Paralimnophyes* from *Limnophyes* is the number of inner mandibular teeth (4 and 3, respectively), therefore, larval remains without mandibles are indistinguishable. Even though the specimens from Central American lake sediments have 4 mandibular inner teeth, indicating *Paralimnophyes*, remains lacking mandibles cannot be distinguished, and could belong to either or both genera. *Paralimnophyes* has not been recorded from Central America yet (Spies et al 2009), with the closest record is from Mexico (Alcocer et al. 1993). *Paralimnophyes* also closely resembles *Compterosmittia*, by the shape of mentum and the 4 inner mandibular teeth. These genera are indistinguishable as a subfossil. Although there are records of *Compterosmittia* from the Neotropics, due to the specific life-strategy of the larvae, associated with pitcher plants, the genus is not likely to be present in lake sediments.

Ecology and Distribution. Limnophyes is a widespread and species-rich genus with described species from Central America. Larvae of most *Limnophyes* species are terrestric/semi-terrestric, only few species are

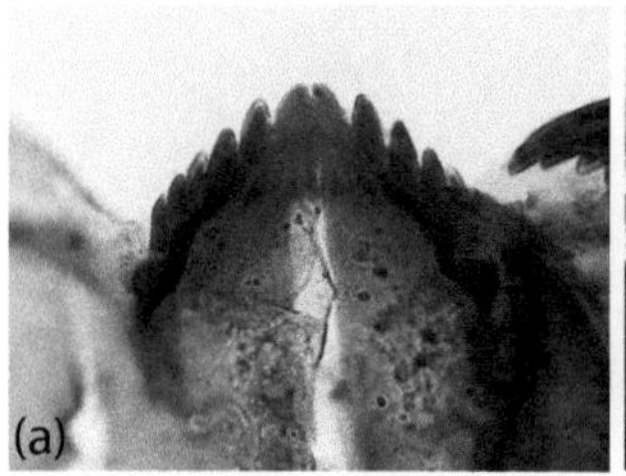

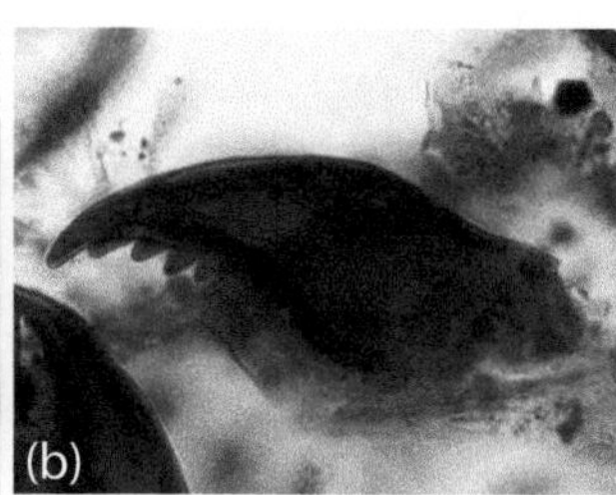

Figure 5.8 *Limnophyes/Paralimnophyes.* a—mentum (with characteristic bulbous lobe at the base), b—mandible, c—premandible

truly aquatic. Larvae can be found in the upper layers of wet soil, hygropetric environments, but also on plants and wood near the water surface of usually smaller stagnant and flowing waterbodies (Moller Pillot 2013). The closely related *Paralimnophyes* live in small lowland waterbodies and bogs at the water table and among vegetation (Moller Pillot 2013). Rare morphotype, recorded in high elevation lakes.

METRIOCNEMUS VAN DER WULP

Diagnosis. Head capsule usually dark. Mentum with 1–2 median teeth and 5–6 lateral; generally, median teeth lower than 1st lateral (Figure 5.9). Premandible with 2–4 apical teeth. Mandible with short apical tooth and 4 inner teeth. Antenna 5-segmented, variable in size from normally developed to reduced.

Remarks. The double median tooth that is smaller than 1st lateral should be characteristic enough to separate the genus. The morphotype recorded, *Metriocnemus eurynotus*-type (following Brooks et al. 2007 and Andersen et al. 2013), has one pair of median teeth slightly smaller than 1st lateral, lateral teeth gradually diminishing in size. Antenna twice as long as broad.

Ecology and Distribution. Metriocnemus is a genus with worldwide distribution, with more than 60 Holarctic species (Cranston et al. 1989), and 5 endemic species from the Neotropical region (Spies and

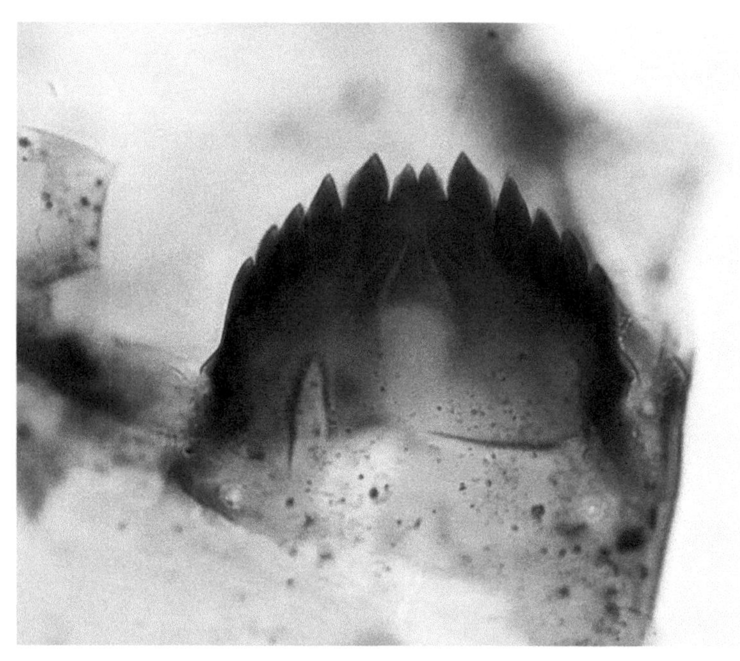

Figure 5.9 *Metriocnemus eurynotus*-type. Mentum.

Reiss 1996, Donato and Paggi 2005). The genus occurs in one of the widest biotope ranges of any dipteran, from mosses and higher vegetation, pitcher plants and hollow trees, to margins of springs, ditches, streams, damp soils, and hygropetric biotopes, lakes and rock pools (Sæther 1989). Rare in Central America, only recorded in high elevation waterbodies.

NANOCLADIUS KIEFFER

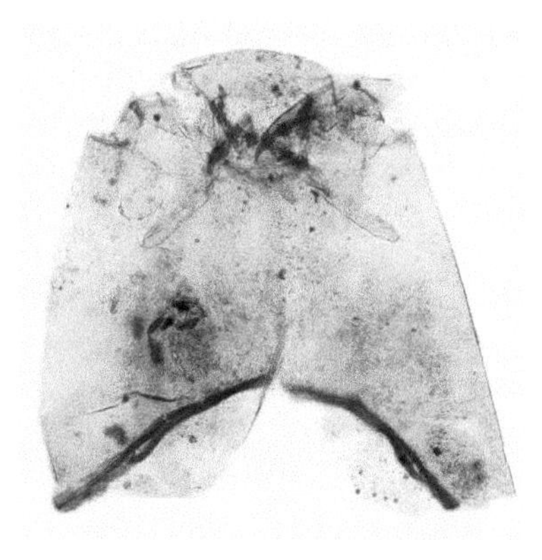

Diagnosis. Distinctive mentum with wide median area bearing two nipple-like median projections, and usually 6 pairs of lateral teeth that are often indistinct and may be fused (Figure 5.10a). Exceptionally large ventromental plates extend beyond the lateral margin of the mentum (Figure 5.10a), beard absent. Premandible with simple or 3–5 very short apical teeth. Mandible with long slender apical tooth and 3 inner teeth (Figure 5.10b). Antenna 5-segmented, 5th segment hair-like.

Remarks. The characteristic shape of mentum with long ventromental plates, extending well beyond the lateral margin of the mentum without beard distinguishes the genus from others. *Nanocladius rectinervis*-type (sensu Brooks et al. 2007), recorded in Central American lakes, is characteristic by the combination of long ventromental plates and well-defined mental teeth.

Ecology and Distribution. Common and widespread genus. Larvae are free living but some species are phoretic or parasitic on aquatic insects. Larvae inhabit lakes, rivers and streams and are known to tolerate high level of organic pollution. Rare in Central America, found in littoral samples of large lakes.

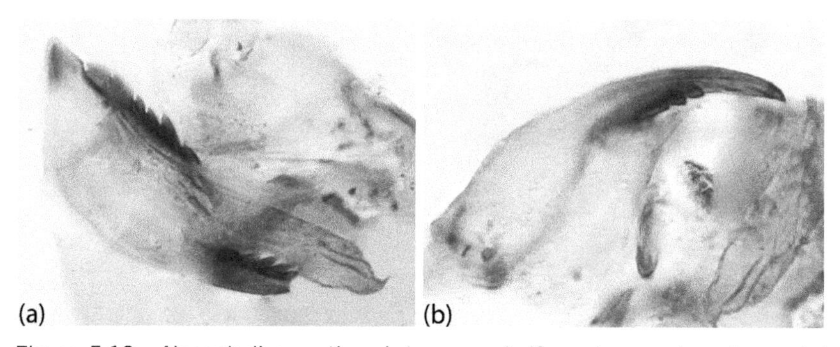

Figure 5.10 *Nanocladius rectinervis*-type. a—half-mentum and ventromental plate, b—mandible.

PARAKIEFFERIELLA THIENEMANN

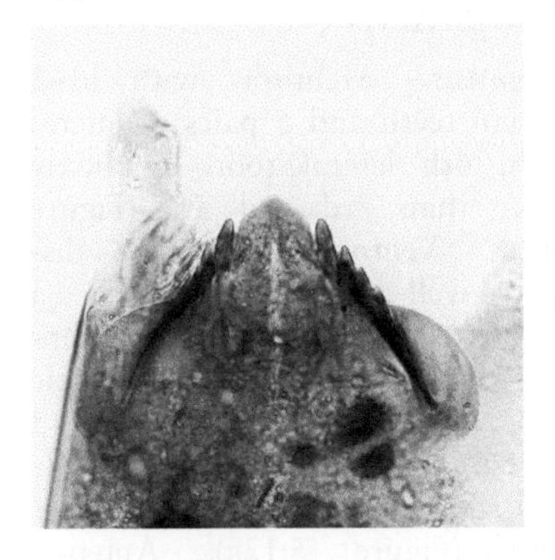

Diagnosis. Mentum with single broad median tooth and 6 lateral teeth (Figure 5.11). Ventromental plates characteristic, strongly developed, broad and protruding beyond the mentum laterally, rounded at apex (Figure 5.11). Premandible with a single or weakly bifid apical tooth. Mandible with a long slender apical tooth and 3 inner teeth. Antenna 6-segmented, 6th segment vestigial.

Remarks. The genus can be distinguished by mentum with an odd number of teeth, well-developed ventromental plates extending beyond lateral margin of the mentum, without a beard. Head capsules from Central American lakes resemble *Parakiefferiella bathophila*-type (following Brooks et al. 2007; Hofmann 1971), but they lack the pale accessory tooth between the median tooth and 1st lateral.

Ecology and Distribution. Genus with worldwide distribution, but rarely recorded from the Neotropics. Epler (2017) reports the genus from Costa Rica without naming the species. Preimaginal stages of *Parakiefferiella* inhabit all kinds of waterbodies from flowing to standing. Rare morphotype associated with high elevation waterbodies.

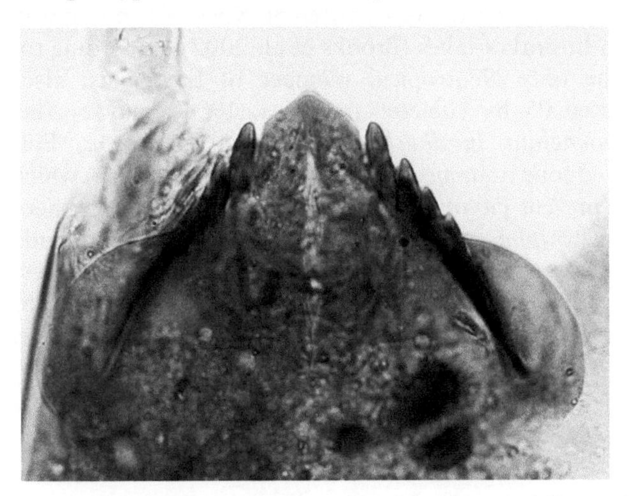

Figure 5.11 *Parakiefferiella.* Mentum with the characteristic ventromental plates.

PARAMETRIOCNEMUS GOETGHEBUER / *PARAPHAENOCLADIUS* THIENEMANN

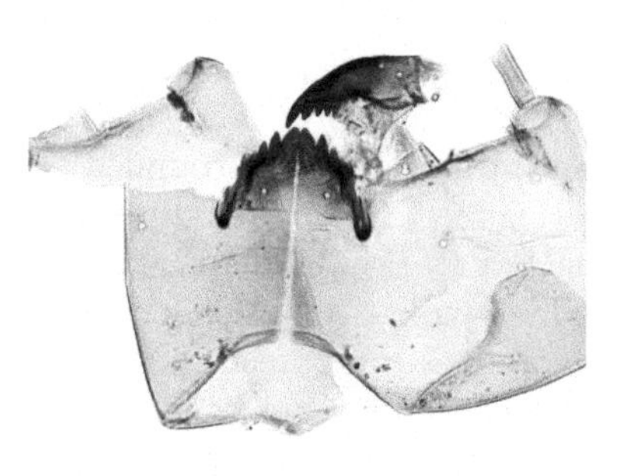

Diagnosis. Mentum with 1–2 median teeth and 5 pairs of lateral teeth, 4th lateral tooth distinctly larger than 3rd and 5th (Figure 5.12a). Ventromental plate distinct, well-developed, forming a lobe extending beyond outer mentum (Figure 5.12a,b). Premandible with 2–6 teeth (however, always 3 in *Paraphaenocladius*). Mandible with short apical tooth and 3 inner teeth (Figure 5.12a). Antenna 5-segmented.

Remarks. The only reliable feature to separate larval *Parametriocnemus* from *Paraphaenocladius* is the curved preanal segment and as a result the posterior direction of procerci and anal setae in *Paraphaenocladius.* Without body, larval *Parametriocnemus* virtually cannot be distinguished from *Paraphaenocladius. Parametriocnemus* may also be confused with *Limnophyes*, but it differs from it by the darkly pigmented lobe below outermost lateral tooth, subequal lateral teeth, and weakly developed ventromental plates.

Ecology and Distribution. Parametriocnemus occurs in all biogeographic regions. Larvae live in different types of running waters, occasionally in littoral of lakes (Brooks et al. 2007). According to Cranston (2010), the only Neotropical member of the genus, also reported from Guatemala by Sublette and Sasa (1994), may be the Nearctic *Parametriocnemus lundbeckii* (Johannsen). However, Epler (2017) documented four unnamed specimens from Costa Rica, which does not seem to represent *Parametriocnemus lundbeckii* (J. Epler pers. comm.). Regarding *Paraphaenocladius*, this is a genus with worldwide distribution, with most species known to be terrestrial and semi-terrestrial, with a few truly aquatic species, inhabiting springs and streams. Sæther and Wang (1995) suggest that South American records of the genus likely belong to the subspecies *Paraphaenocladius exagitans* subsp. *longipes* (Sæther and Wang). Rare in Central American lakes.

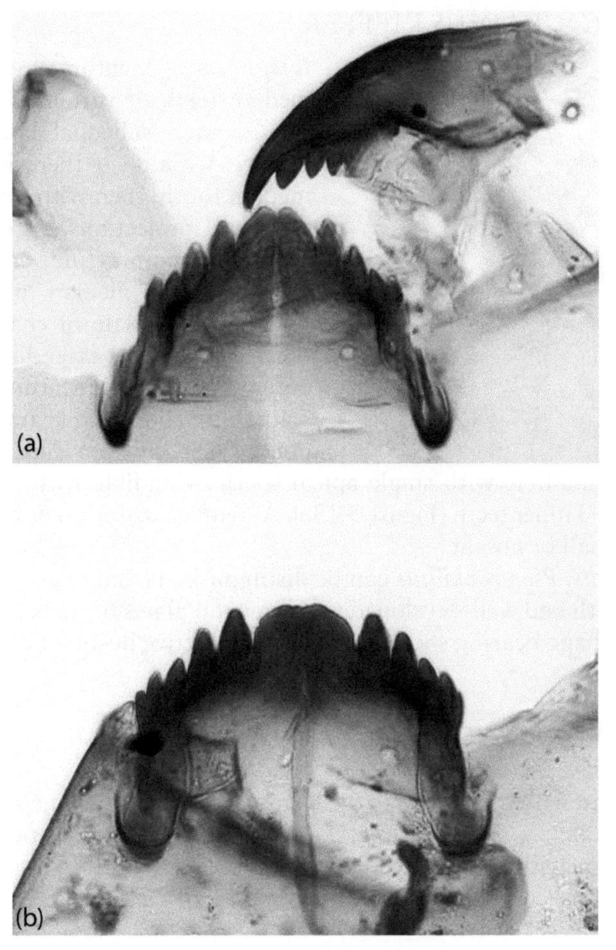

Figure 5.12 *Parametriocnemus/Paraphaenocladius.* a—mentum and mandible, b—mentum (worn down).

PSECTROCLADIUS KIEFFER

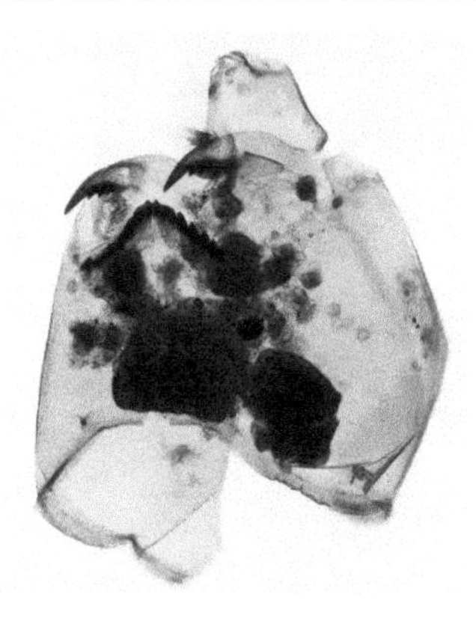

Diagnosis. Mentum with 1–2 median teeth of various shapes and sizes, and 5 subequal lateral teeth (Figure 5.13a,d); if there is a single median tooth, then with low median or lateral projections, with triangular median point, trifid, or with pair of nipple-like median projections. Ventromental plate of characteristic triangular shape, extending beyond outer tooth of mentum (Figure 5.13a,b,d); short and sparse beard is present (usually not visible in subfossils). Premandible with single apical tooth. Mandible with long apical tooth and 3 inner teeth (Figure 5.13e). Antenna 5-segmented, Lauterborn organs small or absent.

Remarks. Psectrocladius can be distinguished from other genera by 5 lateral teeth and well-developed ventromental plates of characteristic triangular shape bearing short beard (not distinctive in subfossil material).

KEY TO MORPHOTYPES

1 Mentum with median teeth well-separated, with short median projection.. *Psectrocladius sordidellus*-type

1' Mentum with median teeth broader, weakly separated with nipple-shaped projections ....*Psectrocladius psilopterus*-type

ADDITIONAL REMARKS TO MORPHOTYPES

Psectrocladius sordidellus-type (Figure 5.13a,b) has mentum with median teeth with a short median projection, lateral teeth subequal and pointed. Ventromental plates characteristically triangular. Beard short and sparse beard.

Psectrocladius psilopterus-type (Figure 5.13c–f) resembles *Psectrocladius sordidellus*-type in shape of mentum, but it has much broader median mental teeth, defined by Cranston et al. (1983), as "single broad median tooth with median nipple-shaped projections".

In subfossil material there is often a great number of intermediate specimens of *Psectrocladius*, thus the two morphotypes are frequently merged. Menta of specimens with buccal deformities may appear as with trifid median tooth.

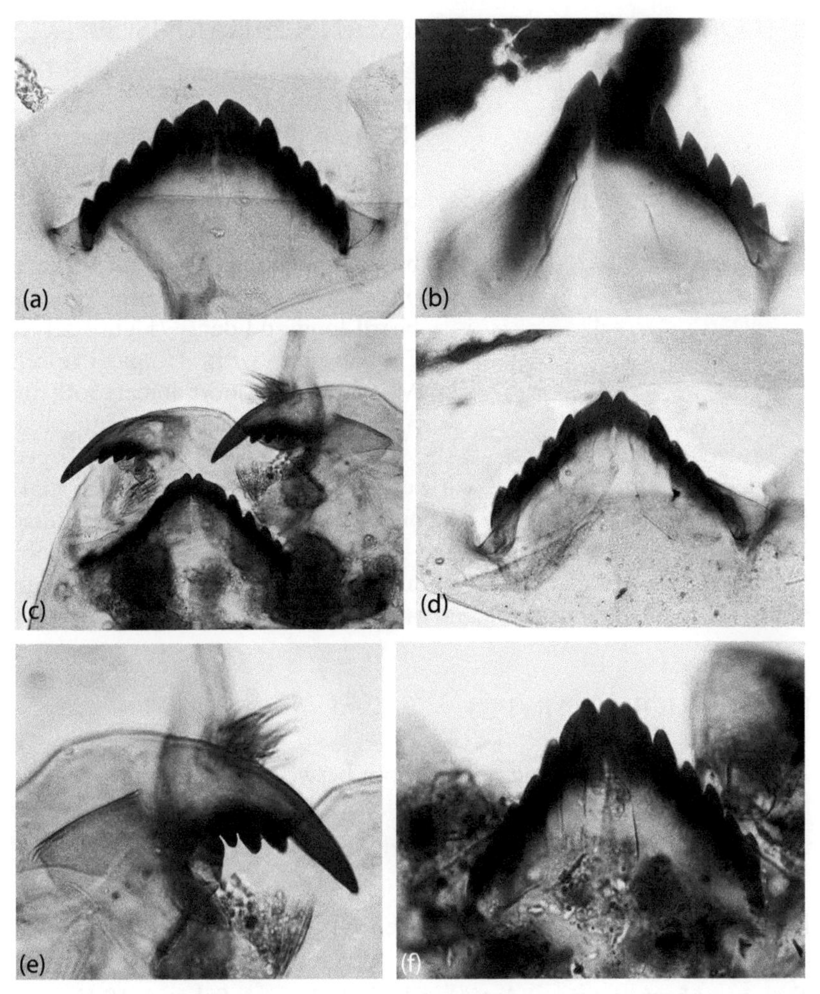

Figure 5.13 *Psectrocladius. Psectrocladius sordidellus*-type: a, b—mentum; *Psectrocladius psilopterus*-type: c—mentum and mandible, d—mentum, e—mandible, f—buccal deformity (trifid median tooth).

Ecology and Distribution. Psectrocladius larvae are lentic, occurring in all kinds of standing waters. The genus is essentially a northern hemisphere taxon.

Together with *Cricotopus* and *Corynoneura*, *Psectrocladius* was among the most common orthoclads found in Central American lakes; larvae show eurizonal distribution, i.e., occurred along the full elevational gradient from lowlands to high elevation lakes. Rare morphotype, only recorded in high elevation lakes.

RHEOCRICOTOPUS THIENEMANN & HARNISCH

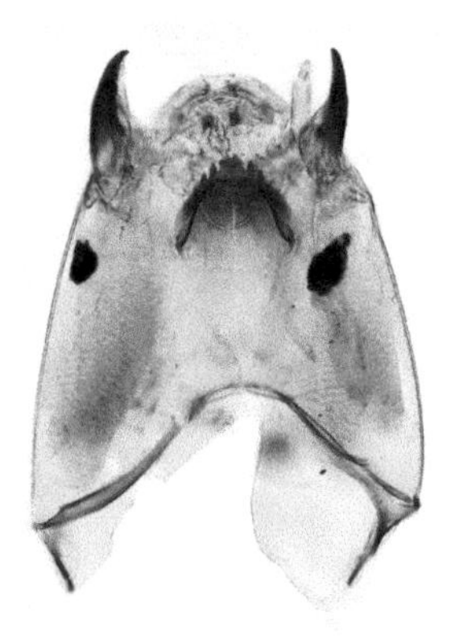

Diagnosis. Mentum with bifid median tooth and 5 lateral teeth (Figure 5.14b); in some species/types there may be a minute accessory tooth between median and 1st lateral tooth (Figure 5.14b). Ventromental plate broad, extending beyond margin of outer mental tooth, apically truncate; beard long and dense (Figure 5.14c). Premandible with 1 apical tooth. Mandible with short apical tooth and 3 inner teeth. Antenna 5-segmented.

Remarks. The strongly developed, bearded ventromental plates with combination of bifid mentum make *Rheocricotopus* unlikely to be confused with other Orthocladiinae. Morphotypes without accessory tooth may superficially resemble *Psectrocladius sordidellus*-type, which has smaller ventromental plates of distinctly triangular shape and only few short setae in beard (if any).

KEY TO MORPHOTYPES

1 Mentum with small accessory tooth between median and 1st lateral teeth...*Rheocricotopus fuscipes*-type

1' Mentum with acessory tooth absent.......*Rheocricotopus effusus*-type

DIAGNOSIS OF MORPHOTYPES AND ADDITIONAL REMARKS

Two morphotypes were distinguished (following Schmid 1993 *apud* Brooks et al. 2007).

Rheocricotopus fuscipes-type (Figure 5.14a) has mentum with a small accessory tooth between the median and 1st lateral teeth in addition to 5 lateral teeth.

Rheocricotopus effusus-type (Figure 5.14b,c) has a mentum lacking the accessory tooth next to the median tooth.

Ecology and Distribution. Larvae of *Rheocricotopus* are in general rheophilic and occur in rivers and streams but can also be found in littoral zone of nutrient-poor lakes. Rare morphotype, only recorded in high elevation lakes.

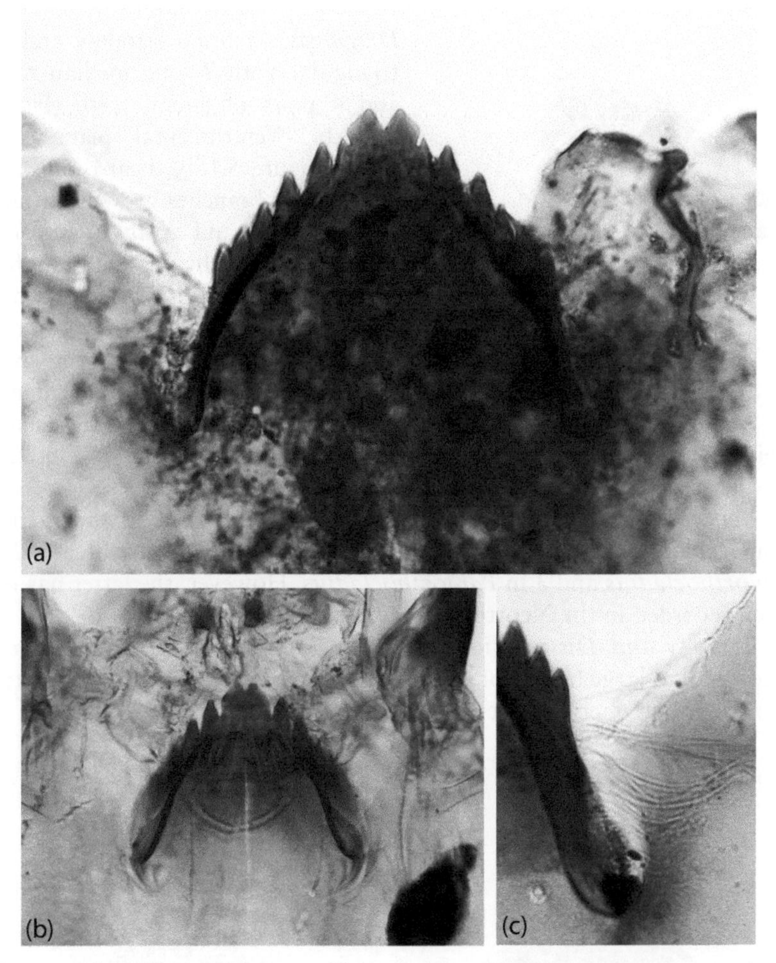

Figure 5.14 *Rheocricotopus. Rheocricotopus fuscipes*-type: a—mentum; *Rheocricotopus effusus*-type: b—mentum, c—beard on ventromental plates.

SYNORTHOCLADIUS THIENEMANN

Diagnosis. Mentum strongly arched, triangular with 2 long median teeth and 4 pairs of lateral teeth (Figure 5.15a,b). Ventromental plate elongated (Figure 5.15b); beard long and dense, with branches growing from a common area and arranged in a circular shape, usually missing in subfossils. Premandible with 1 apical tooth. Mandible with short apical tooth and 3 inner teeth, there is a strong spine on the mola (Figure 5.15a). Antenna 4-segmented; 2nd segment usually apically broadened; 3rd segment may be subdivided into 2 parts, looking like a 5-segmented antenna.

Remarks. Larval *Synorthocladius* resemble *Parorthocladius* from which they can be separated by the number of median mental teeth: 2 in *Synorthocladius* and 3 in *Paraorthocladius*. However, the latter has not been recorded in the Neotropical region.

Ecology and Distribution. Synorthocladius larvae inhabit springs and small to large bodies of standing and flowing water. Rare in Central America, but when it presents it can be abundant.

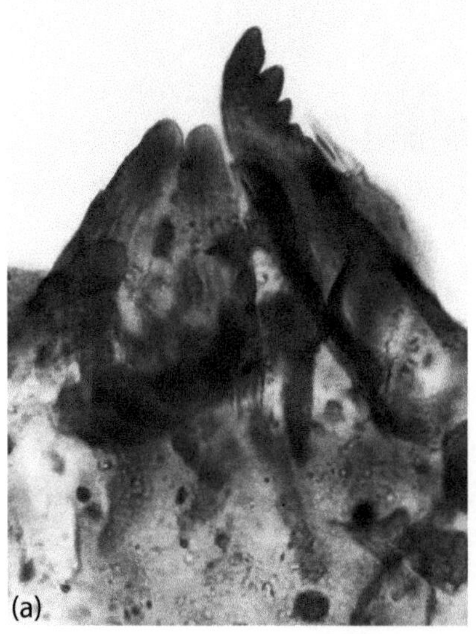

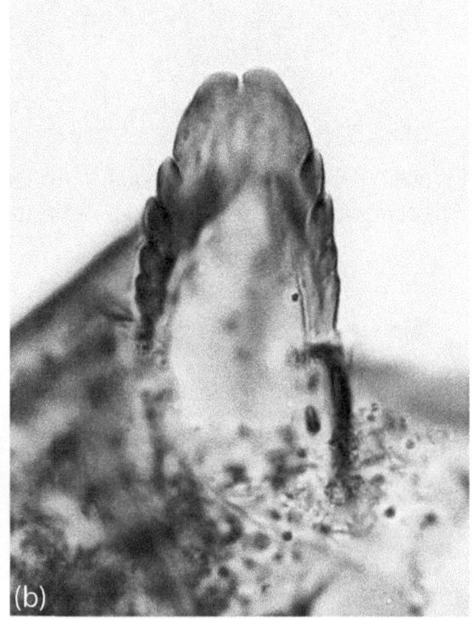

Figure 5.15 *Synorthocladius.* a—mentum and mandible with the strong spine on the mola, b—mentum (worn down; note elongate ventromental plate).

THIENEMANNIELLA KIEFFER

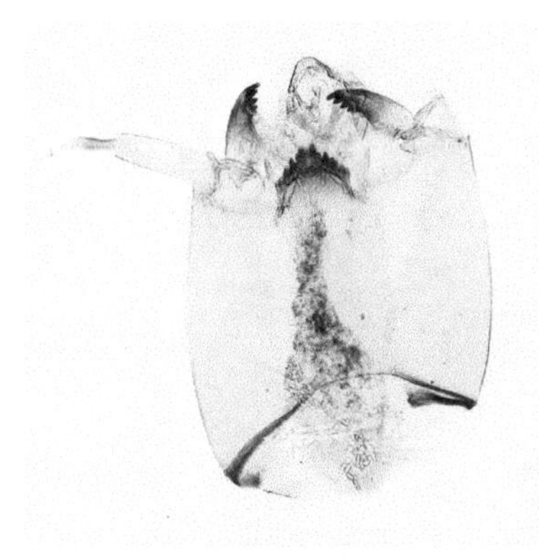

Diagnosis. Head capsule small, prolonged oval, often dark pigmented (Figure 5.16a). Mentum with 2–3 median teeth and 5 pairs of lateral teeth (Figure 5.16b). Premandible with 1 apical tooth. Mandible with apical tooth subequal to 1st of 4 inner teeth (Figure 5.16b). Antenna at least ½ but no more than ¾ length of head; 5-segmented; segment 3 unusually long; segments 4 and 5 small; 3rd segment is frequently darker than the rest of antenna (Figure 5.16c).

Remarks. *Thienemanniella* is similar to *Corynoneura* and *Onconeura* and can be distinguished from them by the proportion of antenna to head: as long or longer than the head in *Corynoneura*; about ⅓ in *Onconeura* and from ½ to ¾ length of the head in *Thienemanniella*. Moreover, *Thienemanniella* heads are usually darker, strongly pigmented, while in *Corynoneura* heads are typically paler. The morphotype found in Central American lakes has a very long 3rd antennal segment, equalling the brown 2nd antennal segment, and the central median tooth is minute, at most about ⅓ the height of the outer median teeth. These characters resemble *Thienemanniella* sp. B sensu Epler (2001).

Ecology and Distribution. Widespread and species-rich genus, named species are known also from Central America (Spies and Reiss 1996; Spies et al. 2009). *Thienemanniella* larvae are found in all kinds of lotic habitats from mountain streams to lowland rivers. Similar to *Corynoneura*, larvae and head capsules of *Thienemanniella* are tiny and easy to overlook. Rare morphotype in Central American lake sediments.

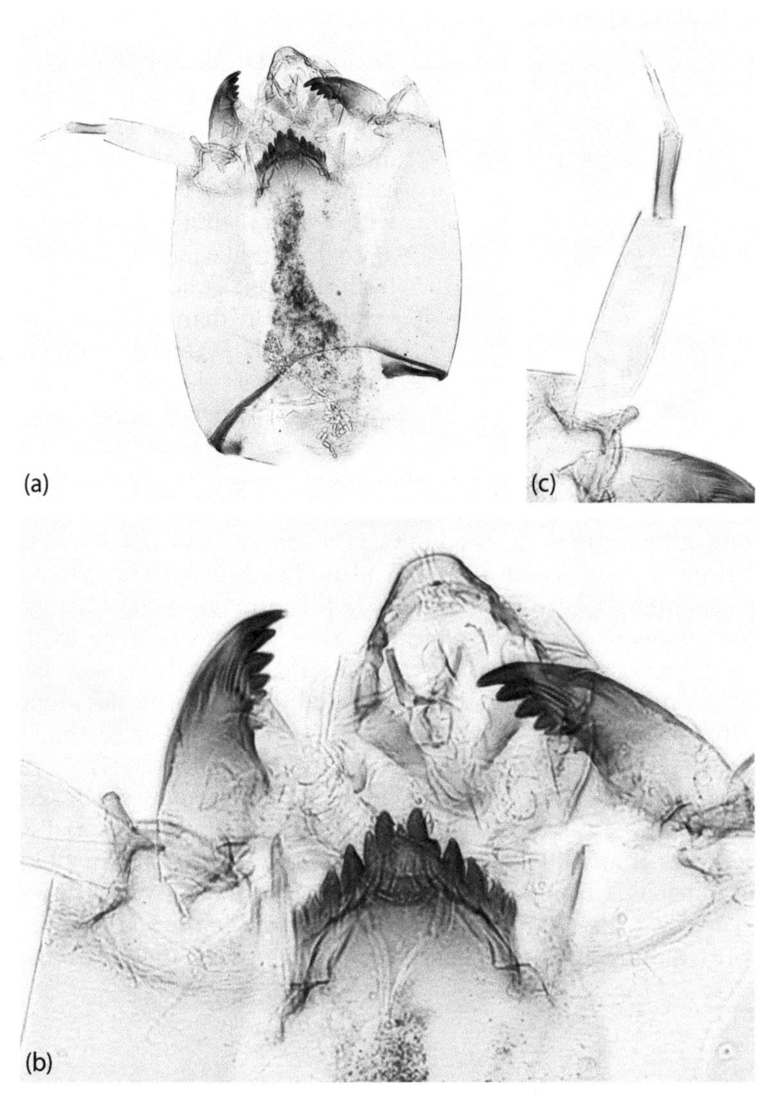

Figure 5.16 *Thienemanniella.* a—head capsule, b—detail of head capsule showing mentum and mandible, c—antenna with darker 2nd segment.

ORTHOCLADIINAE TYPE A

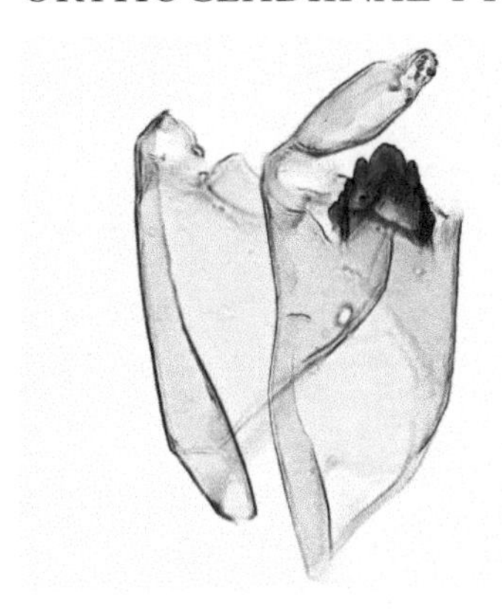

Diagnosis. Mentum with single domed median tooth, and 4 pairs of lateral teeth, outermost lateral tooth notched medially (Figure 5.17). Ventromental plate small, beard absent. Mandible and premandible unknown.

Remarks. This morphotype resembles semi-terrestrial genera, such as *Mesosmittia* and *Bryophaenocladius*. Both are common in the Neotropics and are represented by multiple species (Spies et al. 2009). It also shows some similarity to *Paracricotopus millrockensis* (see Epler 2001), living in seeps, bogs, springs and low order streams. However, *Paracricotopus* has not been recorded in the Neotropics so far.

Ecology and Distribution. Rare morphotype, found only in one sediment core taken from lake Apastepeque, El Salvador.

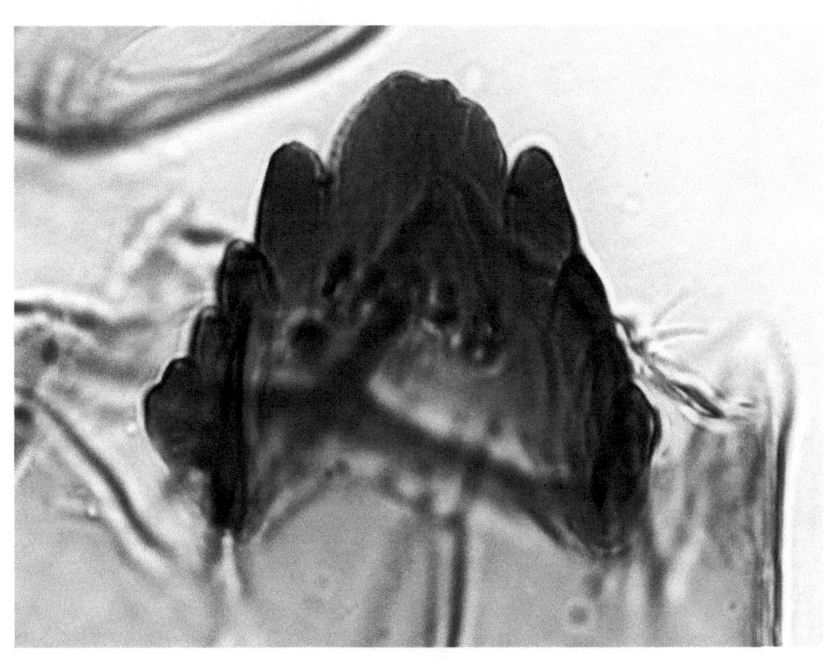

Figure 5.17 Orthocladiinae type A Mentum.

6 Subfamily Podonominae

Larval Podonominae are characterized by the absence of premandible and weak ventromental plates. The shape of mentum is typical and consists of a single median tooth and 7–15 pairs of lateral teeth. Outer margin of mandible is strongly bent in the middle, with a large apical, a small outer subapical, and 4–9 inner teeth.

DOI: 10.1201/9780429021572-6

PAROCHLUS ENDERLEIN

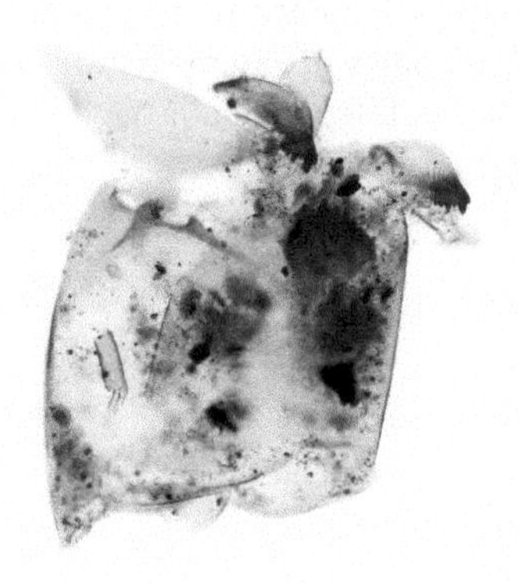

Diagnosis. Head capsule small to medium size, light brown. Mentum with 1 median tooth and 7 lateral teeth, the first two lateral teeth are shorter than the 3rd one; outer lateral teeth gradually diminishing in size (Figure 6.1a). Premandible absent. Mandibles with 5 teeth on the inner margin, apical and 2nd inner tooth are the largest, 1st inner tooth is minute, and with 1 dorsal tooth (Figure 6.1b). Antenna 4-segmented (Figure 6.1c), if 5-segmented, 3rd segment annulate.

Remarks. The shape of mentum and mandible of *Parochlus* is unique and will differentiate the genus from other chironomids.

Ecology and Distribution. *Parochlus* is a species-rich, cold-stenothermic genus with a wide ecological valence in running waters of high latitudes and elevations. Larvae are mainly associated with different types of high elevation streams, but several species prefer springs and some occur among mosses in shallow tarns and other small waterbodies (Brundin 1983). The genus is almost completely confined to the southern temperate zones, including the Andean mountain chains and the South Shetland Islands in true Antarctic latitudes. Rare in Central American lakes, only occurring in high elevations.

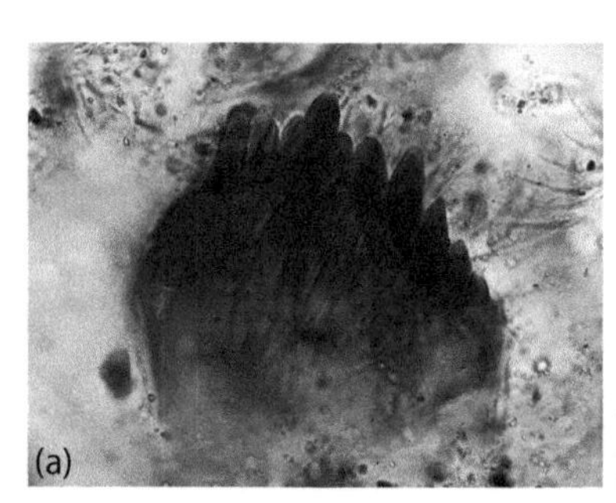

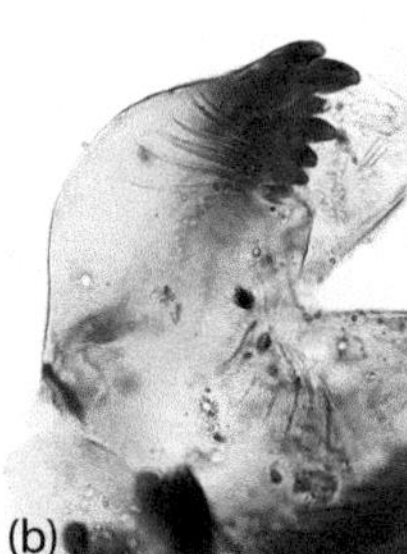

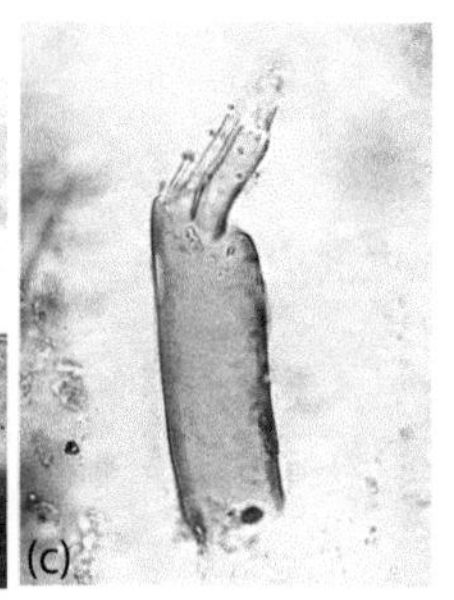

Figure 6.1 *Parochlus*. a—mentum, b—mandible, c—antenna.

The typical predatory life style of Tanypodinae is reflected in the larval mouth apparatus, which differs from other Chironomidae by having a strong development of premental structures such as the ligula and paraligula (Cranston 1995). Additionally, the group is easily recognisable by its antenna, capable of complete retraction into head, a feature that is unique in Chironomidae. In subfossil material, however, important diagnostic structures may be missing, which makes even generic identification difficult. In such cases, the pattern of cephalic setation and sensory pores can be used for identification (see Kowalyk 1985; Rieradevall and Brooks 2001; Brooks et al. 2007; Cranston and Epler 2013). Figure 7.1 shows the typical features of subfossil larval Tanypodinae.

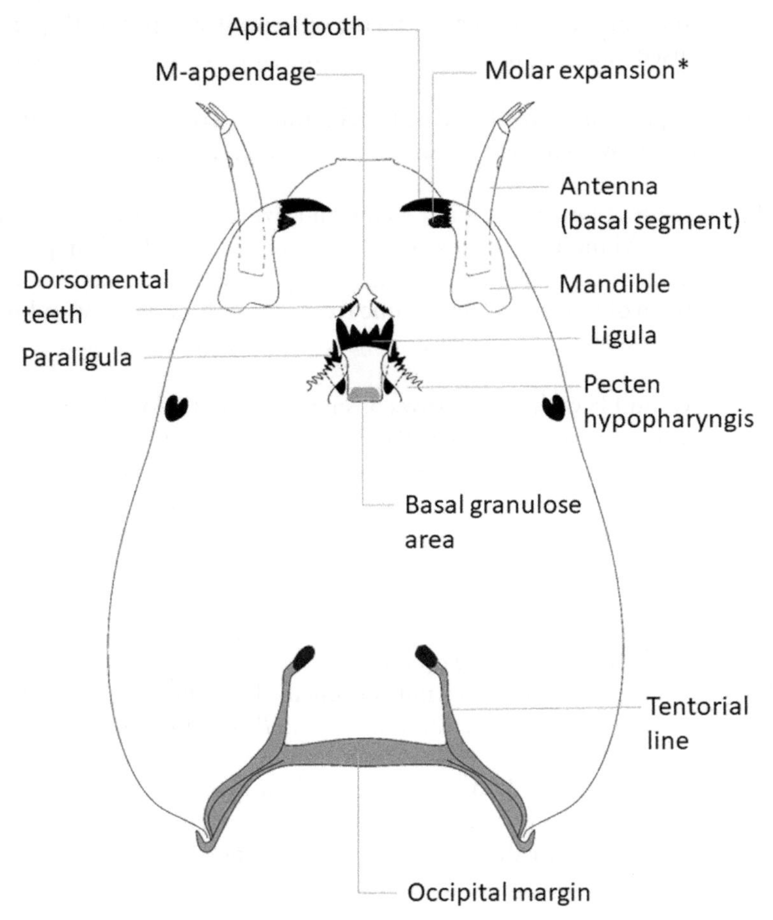

Figure 7.1 Tanypodinae larval head capsule showing subfossil characters. *In older literature it is called "basal tooth" (Fittkau and Roback 1983).

DOI: 10.1201/9780429021572-7

KEY TO GENERA OF TANYPODINAE*

1 Head rounded, dorsomental teeth in well-developed transverse plates....2
1' Head elongated, dorsomental teeth absent....8

2(1) Ligula with 6–7 teeth (occasionally more)....3
2' Ligula with 4–5 teeth....4

3(2) Mandible with strongly hooked apical tooth and large inner tooth, ligula with even number of teeth (usually 6). **Tribe Clinotanypodini** .... *Clinotanypus*
3' Mandible with slightly curved apical tooth, inner tooth small and curvy; ligula with 7 teeth, outermost 2 inner teeth partly fused .... *Coelotanypus*

4(2) Ligula with 4 teeth. **Tribe Procladiini** in part.... *Djalmabatista*
4' Ligula with 5 teeth....5

5(4) Teeth of ligula of similar shape and size, pale, row of teeth convex. Mandible, if present, with stout base. **Tribe Tanypodini** .... *Tanypus*
5' Teeth of ligula straight or concave, ligula pale or dark. Mandible, if present, more elongate....6

6(5) Mandible with several rows of small teeth both dorsally and ventrally. Ligula with inner teeth appressed to median tooth; dorsomental teeth arranged in concave arch. **Tribe Fittkauimyiini** .... *Fittkauimyia*
6' Mandible without such row of teeth. Ligula with straight or outwardly curved teeth....7

7(6) Head strongly sclerotized, reddish-brown; paraligula with 1 inner tooth. **Tribe Macropelopiini**....Macropelopiini type A
7' Head pale; paraligula multitoothed. **Tribe Procladiini** in part .... *Procladius/Djalmabatista*

8(1) Ligula with median tooth longer than inner teeth, reaching or exceeding level of outer teeth. **Tribe Pentaneurini**....9
8' Ligula with median tooth equal to or shorter than inner teeth 10

9(8) Head usually with lateral spines or extensively covered with small spines/granules, mandible with large molar expansion .... *Labrundinia*

9' Head smooth, never with lateral spines, or granulation, mandible with small molar expansion .... *Nilotanypus*

10(8) Ligula with tooth row almost straight, teeth of about the same size and height, basal granulose area large............... *Pentaneura*

10' Ligula with tooth row concave... 11

11(10) Basal maxillary palp, when not missing, undivided. Ligula with tooth row slightly concave, basal granulose area narrow stripe, distal half of ligula dark brown .... *Larsia*

11' Basal maxillary palp, when not missing, subdivided into 2–6 segments. Ligula with tooth row moderately to strongly concave, apical ⅓ dark... *Ablabesmyia*

* Modified after Cranston and Epler (2013).

ABLABESMYIA JOHANNSEN

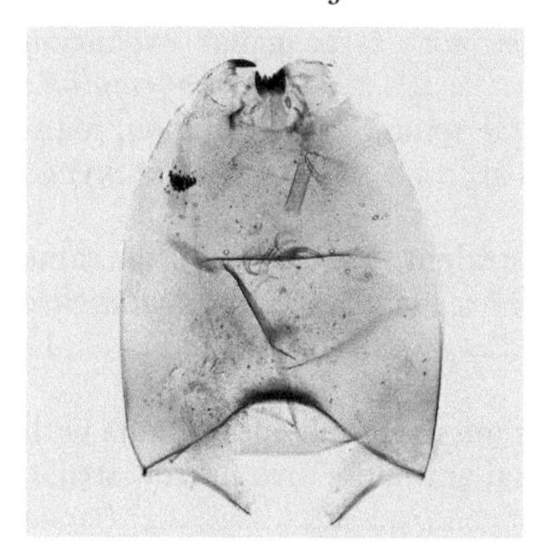

Diagnosis. Head capsule long, narrow, pale yellow-brown with darker occipital margin with or without thorn-like projections on the ventral side (Figure 7.2a,b); cephalic index 0.60–0.65. Maxillary palp basally subdivided into 2–6 segments (3-segmented in the Central American subfossil material, Figure 7.2e). Ligula concave, with 5 teeth (Figure 7.2c,d,g). Paraligula bifid (Figure 7.2c). Pecten hypopharyngis unequal-sized with 12–20 teeth (in 4th instar, Figure 7.2c,d). Mandible with long apical tooth (3× of the basal width), apical half dark brown to black (Figure 7.2f); inner tooth large, blunt, mola expanded to large pointed tooth. Dorsomental teeth absent.

Remarks. Living larvae can be easily distinguished from other tanypods by the maxillary palp subdivided into 2–6 segments, however, this trait is usually absent in subfossil material. According to Vallenduuk and Lipinski (2009), thorn-like projections on the ventral occipital margin will separate *Ablabesmyia* from other Tanypodinae (except for *Zavrelimyia*) but this character was not present in all *Ablabesmyia* remains in the Central American material.

KEY TO MORPHOTYPES

1 Ligula with teeth forming a concave toothed margin .... *Ablabesmyia monilis*-type

1' Ligula with teeth even or, if central tooth smaller, inner teeth subequal to or slightly exceed outer teeth...................... *Ablabesmyia janta*-type

ADDITIONAL REMARKS TO MORPHOTYPES

Two morphotypes were identified following Epler (2001). Both morphotypes have the basal segment of maxillary palp divided into 3 segments (indicating that they belong to the subgenus *Ablabesmyia*) and differ in shape of ligula.

Ablabesmyia monilis-type (Figure 7.2c,d) ligula with teeth increasing in size from the middle tooth to the outer teeth, forming a concave toothed margin.

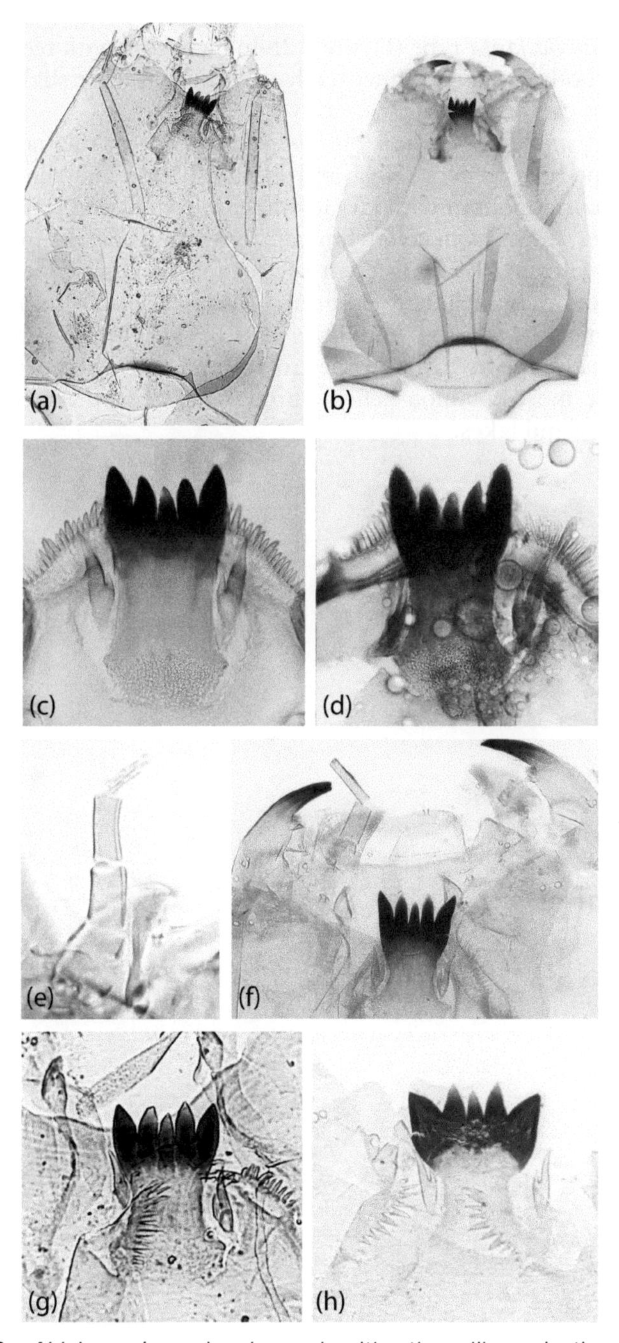

Figure 7.2 *Ablabesmyia*. a—head capsule with a thorn-like projection on occipital margin, b—head capsule without a thorn-like projection on occipital margin; *Ablabesmyia monilis*-type: c, d—ligula, paraligula and pecten hypopharyngis, e—maxillary palp; *Ablabesmyia janta*-type: f—detail of head with ligula, mandible, maxillary palp (divided), g—ligula, h—ligula with variable shape.

Ablabesmyia janta-type (Figure 7.2f–h) has ligula with teeth even or, if central tooth is smaller, inner teeth are subequal to or slightly exceed outer teeth.

Ecology and Distribution. Ablabesmyia is one of the most species-rich Tanypodinae genera (Silva and Ekrem 2016). The genus is eurytopic and cosmopolitan with larvae inhabiting small and large standing and flowing waters from cold temperate to warm tropical climatic zones (Silva and Farrell 2017). Different species may occupy wide variety of habitats, and may include tolerant and sensitive species to acidity and humic content (Cranston and Epler 2013). The highest diversity of *Ablabesmyia* appears to be in the tropics and warm temperate zones. One of the most common genera throughout Central America, especially in lowland lakes.

CLINOTANYPUS KIEFFER

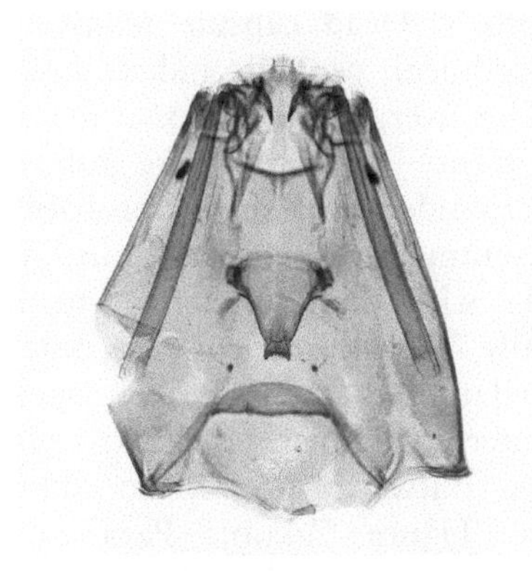

Diagnosis. Head capsule long, conical; cephalic index about 0.7. Ligula with 5–7 teeth, tooth row deeply concave with large outer teeth (Figure 7.3a). Paraligula with 4–5 short lateral teeth. Pecten hypopharyngis elongate, with a dense row of 20–35 teeth (Figure 7.3a). Mandible with distal ⅓ strongly hooked, apical tooth robust, up to thrice as long as basal width; molar expansion large (Figure 7.3a). Dorsomental teeth consist of 12–20 spinules in simple or laterally double row (Figure 7.3b).

Remarks. Larvae of the closely related genera *Clinotanypus* and *Coelotanypus* belong to the tribe Clinotanypodini and are recognized by dorsomental teeth arranged in longitudinal rows on the M-appendage and ligula with 6–7 teeth (Silva and Ekrem 2016). *Clinotanypus* can be separated from *Coelotanypus* by the hooked mandible with a large inner tooth in contrast to the gently curved mandible with a broad, bluntly rounded lamella of *Coelotanypus*. In addition, ligula of *Clinotanypus* usually possesses an even number of teeth (6), while that of *Coelotanypus* an odd number of teeth (7), however, this character has to be considered with caution because both genera have known specimens with both odd and even numbers of teeth on ligula.

Ecology and Distribution. *Clinotanypus* is a widespread genus with larvae inhabiting soft sediments of shallow, warm waterbodies of all sizes, including ponds, lakes and slowly flowing streams and rivers, and of varying water quality (Cranston and Epler 2013). Rare in Central American material, only recorded in lowland lakes.

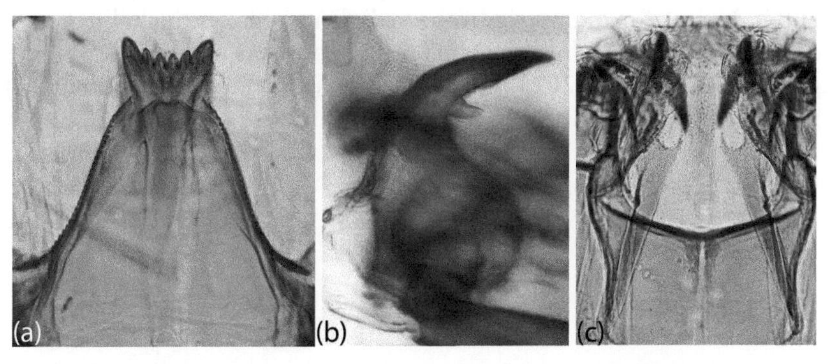

Figure 7.3 *Clinotanypus*. a—ligula with long pecten hypopharyngis on each side, b—mentum, c—mandible (modified after Bitušík and Hamerlík 2014).

COELOTANYPUS KIEFFER

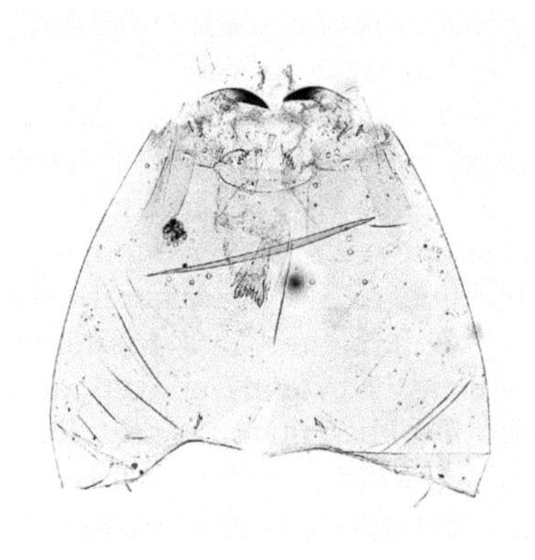

Diagnosis. Head capsule relatively long, conical; cephalic index about 0.7. However, in the subfossil material, plenty of heads were robust with rounded-oval shape, probably representing a younger larval instar. Ligula with odd number of teeth (usually 7), tooth row concave, outer tooth larger than inner ones, apex of outermost inner tooth strongly curved outward, pressed to outer tooth (Figure 7.4a,b). Paraligula strong, elongate with 2–3 spines on each side (Figure 7.4a,b). Pecten hypopharyngis with a narrow row of 18–28 teeth (Figure 7.4a). Mandible curved with slender apical tooth, about thrice as long as basal width, mola expanded into smooth, rounded dorsal lamella (Figure 7.4c). Dorsomentum with coarse granulation extending laterally over a row of 5–10 teeth (Figure 7.4d,e).

Remarks. Coelotanypus can be distinguished from *Clinotanypus* by the gently curved mandible with a broad, bluntly rounded lamella, contrary to the hooked mandible with a large inner tooth of *Clinotanypus*. Even though *Coelotanypus* larvae usually have a ligula with an odd number of teeth, specimens with an even number of teeth are recorded frequently (Epler 2001). Only one morphotype was recorded in Central American lakes: *Coelotanypus concinnus*-type, which possesses 5 dorsomental teeth (for more details on the taxonomy of Clinotanypodini see Roback 1976). However, a mentum representing most likely a different morphotype was also collected (Figure 7.3d).

Ecology and Distribution. It is a species-rich genus with at least 8 species described from Central America (Spies and Reiss 1996; Andersen et al. 2000) and with many more expected in the region (Spies et al. 2009). Larval *Coelotanypus* live in sediments of lakes, including artificial impoundments, slowly flowing reaches of rivers and in old riverbeds (Cranston and Epler 2013). *Coelotanypus* is a common and abundant genus in lake sediments of Central America.

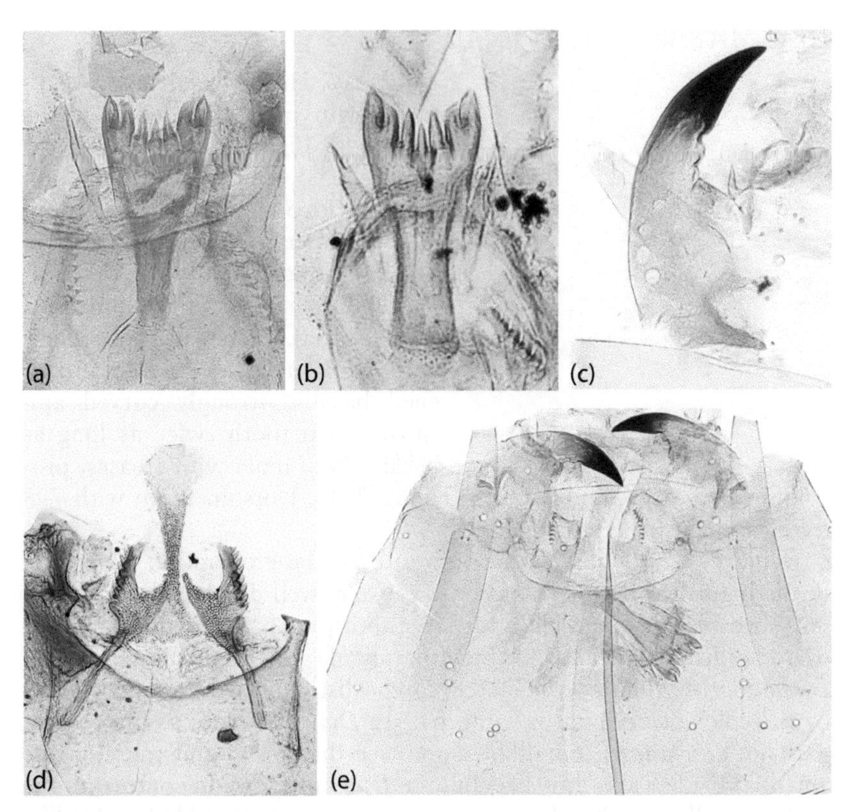

Figure 7.4 *Coelotanypus.* a, b—different types of ligula, c—mandible, d—mentum with dorsomental teeth, e—detail of head showing cephalic setation.

DJALMABATISTA FITTKAU

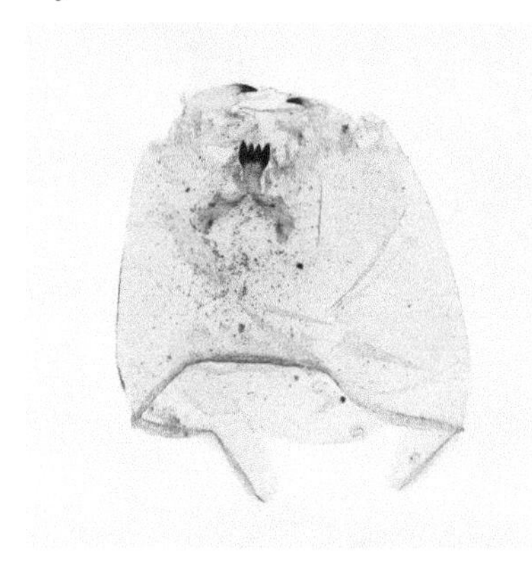

Diagnosis. Head capsule and occipital margin pale, rounded, cephalic index 0.85. Ligula usually with 4 teeth (Figure 7.5a), some species with 5; in species with 5 teeth the median tooth is smaller than inner teeth (Figure 7.12f,g). Apical half of teeth dark. Paraligula short, multi-branched (Figure 7.5a). Pecten hypopharyngis with 10–12 teeth. Mandible broadened basally, strongly curved apically; apical tooth twice as long as basal width; mola with creases, protruding as a large double tooth (Figure 7.5b). Dorsomentum with 7–8 teeth on each side (Figure 7.5a).

Remarks. Djalmabatista belongs to the tribe Procladiini, which can be distinguished by the round head capsule, well-developed dorsomental tooth plates, mandible with large tooth on mola, ligula darkened over the distal half and pectinate paraligula (Silva and Ekrem 2016). *Djalmabatista* (ligula with 5 teeth) can be distinguished from *Procladius* by the color of occipital margin, pale in *Djalmabatista*, as opposed to dark in *Procladius*, mandible shape (see diagnosis) and the elongate antennal blade, exceeding flagellum in *Djalmabatista*, in contrast to the shoter flagellum in *Procladius*. In cases where antennal blade is hyaline and missing, separation of the two genera is problematic.

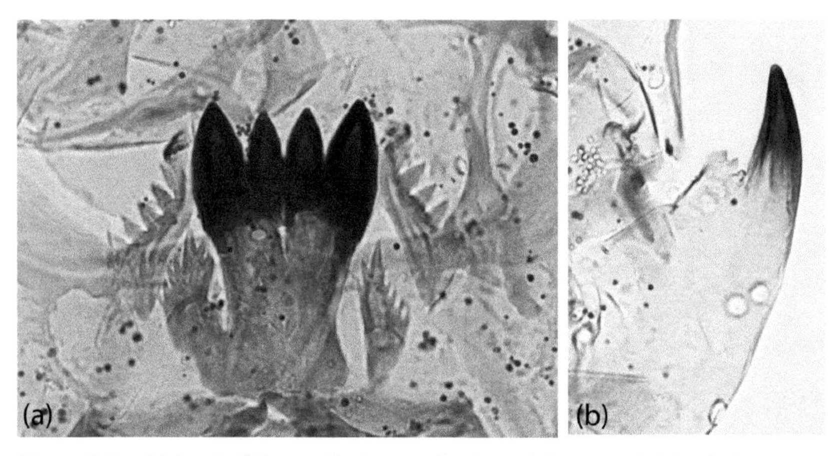

Figure 7.5 *Djalmabatista.* a—ligula, paraligula and dorsomental teeth, b—mandible with protrusion of mola as a large double tooth.

Ecology and Distribution. The genus is diverse in the Neotropical region (Spies et al. 2009). Not much is known about the ecology of the genus *Djalmabatista*, however, *Djalmabatista pulchra*, a widespread species also known from Central America (Spies et al. 2009), was recorded living in soft water or weakly acidic lakes and ponds and in slow flowing rivers in the eastern Nearctic from Canada to Florida (Cranston and Epler 2013). Rare in Central America, recorded only in some lowland lakes.

FITTKAUIMYIA KARUNAKARAN

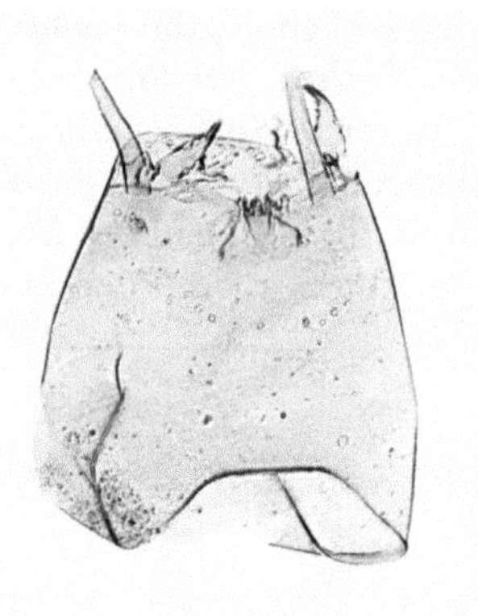

Diagnosis. Head capsule oval, medium-sized to large, cephalic index 0.64–0.75. Ligula with 5 teeth, basal ⅓ strongly narrowed; median tooth shorter and narrower than the remaining teeth; apices of inner teeth pointing strongly toward median tooth (Figure 7.6a). Paraligula consists of a longer inner tooth and 2–4 outer spines (Figure 7.6a). Pecten hypopharyngis consists of up to 22 teeth of laterally decreasing size. Mandible moderately curved; apical tooth indistinctly defined, about ¼ of length of mandible; mola not expanding, with many small dorsal and ventral teeth (Figure 7.6b). Dorsomentum tripartite concave, continuous with lateral complex of structures, comprising an inwardly and an outwardly directed tooth (Figure 7.6a).

Remarks. Fittkauimyia belongs to the monotypic tribe Fittkauimyiini and differs significantly from all other larval tanypods in the arrangement

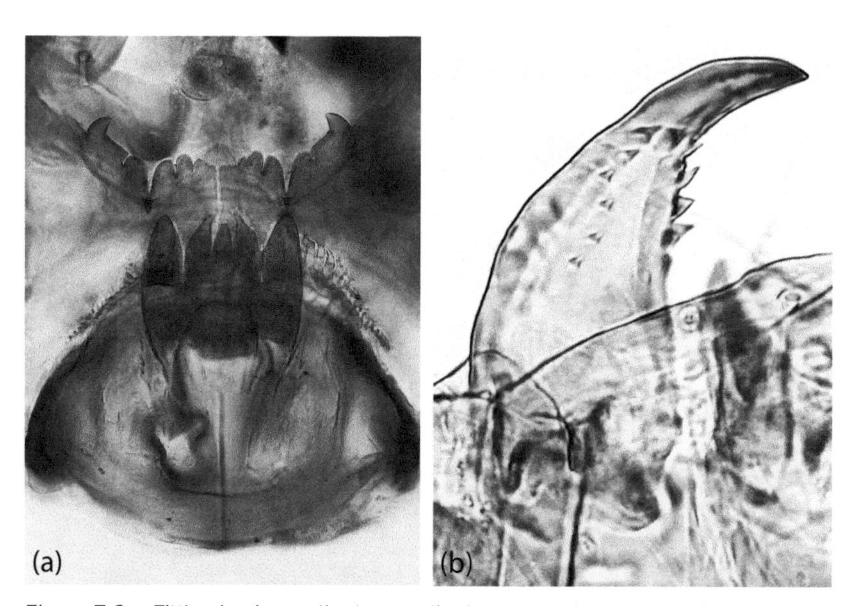

Figure 7.6 *Fittkauimyia.* a—ligula, paraligula, pecten hypopharyngis and dorsomental teeth, b—mandible with small teeth arranged in arch-shape.

of the dorsomental teeth and the shape of the mandible and ligula (Silva and Ekrem 2016).

Ecology and Distribution. This is a cosmopolitan genus characteristic especially for tropical and subtropical regions. Larvae inhabit rivers and the littoral region of lakes. The group was rare in sediments of Central American lowland lakes.

LABRUNDINIA FITTKAU

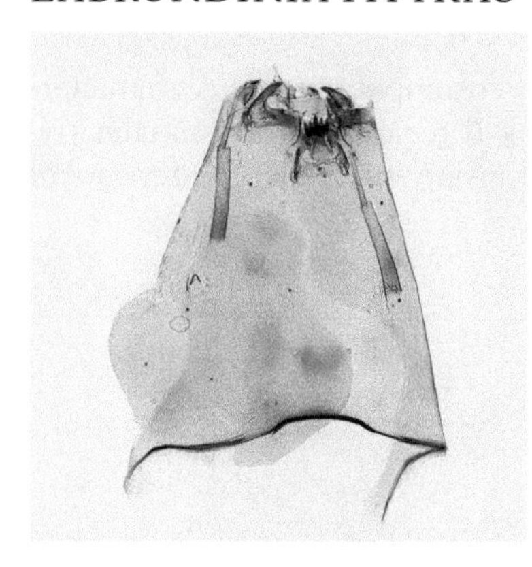

Diagnosis. Head evenly colored, sometimes with ventral maculation; surface may be smooth or covered with spinules (Figure 7.7a), lateroventral and posteroventral spine groups usually present (Figure 7.7b), in some species absent. Ligula with 5 teeth, middle tooth usually longer than outer teeth (Figure 7.7f,g). Paraligula bifid or multitoothed. Pecten hypopharyngis with seven teeth of equal length. Mandible curved, apical tooth long, inner tooth large, molar extension large, robust (Figure 7.7f,g).

Remarks. Head capsules of *Labrundinia* and *Nilotanypus* are similar in the shape of ligula and granulose basal area, but can be separated by a much smaller molar expansion in *Nilotanypus* compared to that in *Labrundinia*. Moreover, several *Labrundinia* species possess a lateroventral or posteroventral group of spines. Remains without spines are difficult to distinguish from *Nilotanypus* subfossils.

KEY TO MORPHOTYPES

1 Surface of head extensively granulated, in well-sclerotized specimens there is a conspicuous Y-shaped mark in the submentum .... *Labrundinia* type Y

1' Head capsule is not extensively granulated and without marking, with lateroventral spine group on each side 2

2 Head with lateroventral spine group present with one single spine on each side 3

2' Head with lateroventral spine group of several spines/spinules on each side 4

3 Head with lateroventral spine group with one single, well developed spine .... *Labrundinia paulae*-type

3' Head with lateroventral spine group with one small and triangular spine .... *Labrundinia* type G

4 Head with lateroventral spine group with rounded spines of variable size, ligula with prominent median tooth, much longer than the lateral teeth .... *Labrundinia virescens*-type

4' Head with lateroventral spine group with few minute spines of triangular shape.. *Labrundinia jasoni*-type

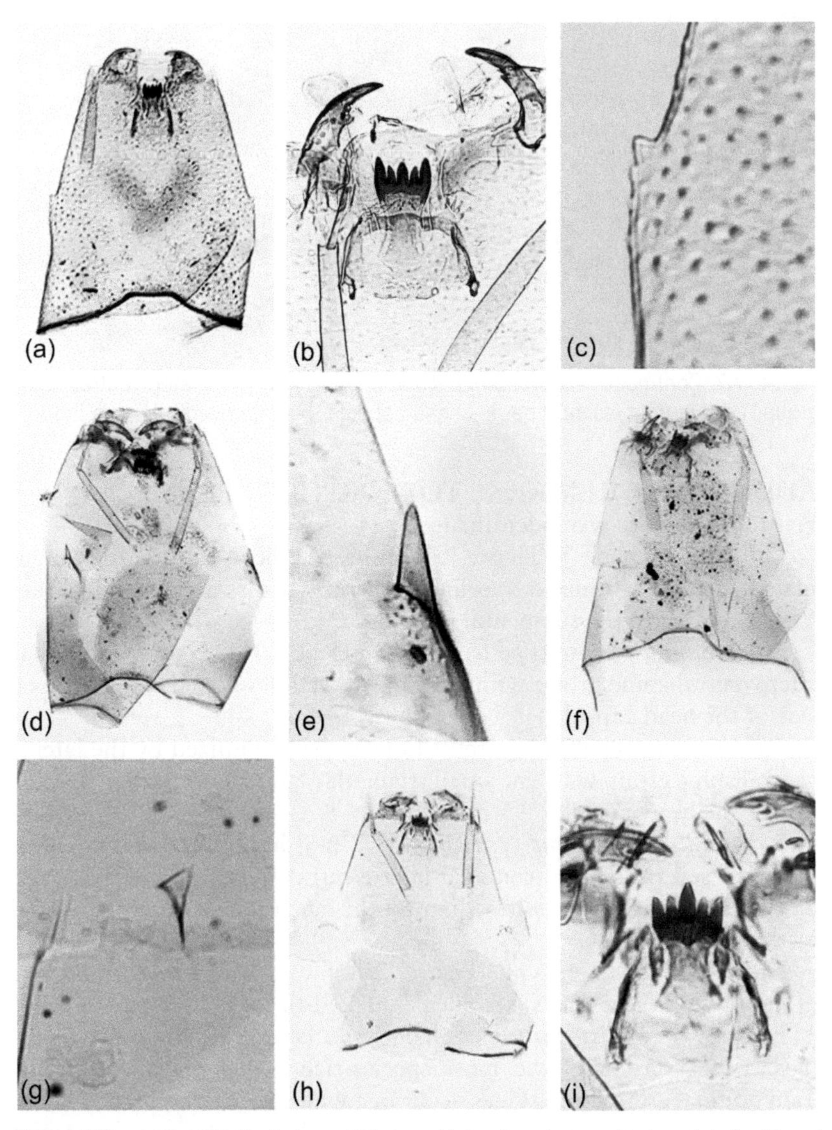

Figure 7.7 *Labrundinia. Labrundinia* type Y: a—head capsule, b—detail of head capsule, c—lateroventral spine group; *Labrundinia paulae*-type: d—head capsule, e—lateroventral spine group; *Labrundinia* type G: f—head capsule, g—lateroventral spine group; *Labrundinia virescens*-type: h—head capsule, i—detail of head.

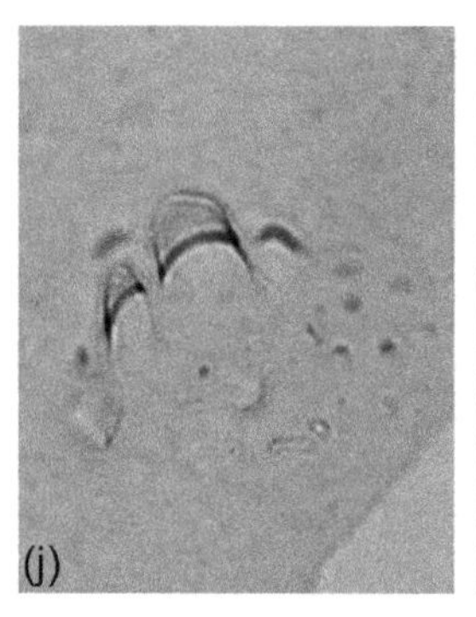

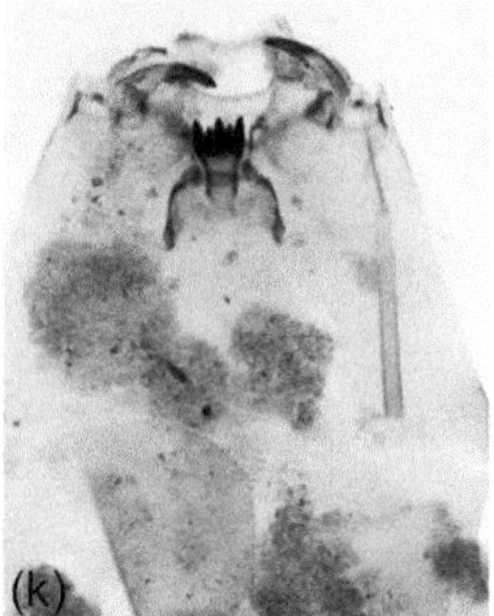

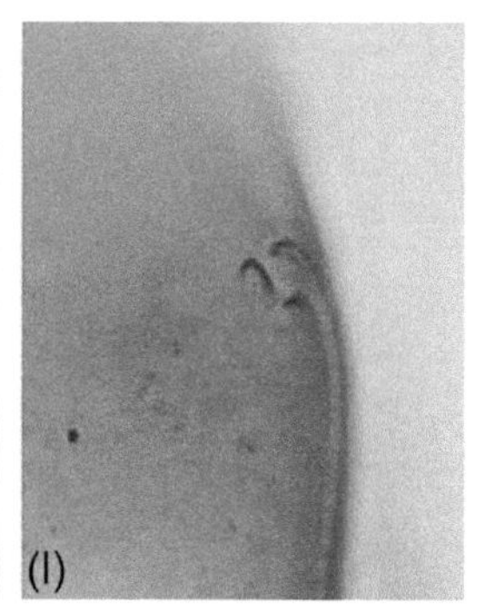

Figure 7.7 Continued *Labrundinia. Labrundinia* type Y: j—lateroventral spine group; *Labrundinia jasoni*-type: k—head capsule, l— lateroventral spine group.

ADDITIONAL REMARKS TO MORPHOTYPES

Five morphotypes were identified.

Labrundinia type Y (Figure 7.7a–c) has the head capsule fully crenulated, in well-chitinized specimens there is a distinct ventral darker Y-shaped area in the submental region.

Labrundinia paulae-type (sensu Silva et al. 2014, Figure 7.7d,e) has lateroventral spine group with one single, well developed spine on each side of the head capsule.

Labrundinia type G (Figure 7.7f,g) can be recognized by the lateroventral spine group with one small triangular lateroventral spine on each side of the head capsule.

Labrundinia virescens-type (Figure 7.7h–j) has a lateroventral spine group of several strong, rounded lateroventral spines of variable sizes, and ligula with prominent median tooth, much longer than the lateral teeth.

Labrundinia jasoni-type (Figure 7.7k,l) has a lateroventral spine group of few weak spines (see Silva et al. 2014).

Ecology and Distribution. Labrundinia is a large, mainly Neotropical genus, probably the most species-rich genus of the subfamily Tanypodinae. Immature stages occur in a wide range of aquatic systems, from small streams and ponds to large lakes. Larvae are associated with aquatic macrophytes or riparian vegetation in slow flowing streams or rivers (Silva et al. 2011, 2014). According to Silva et al. (2015a), the genus may have had its initial diversification in warmer waters in the Neotropical region and that its current presence in the Nearctic region and southern South America is due to subsequent dispersal. The group is one of the most frequent genera in Central American lake sediments, especially *Labrundinia* type Y and *Labrundinia virescens*-type, recorded all over the study region.

LARSIA FITTKAU

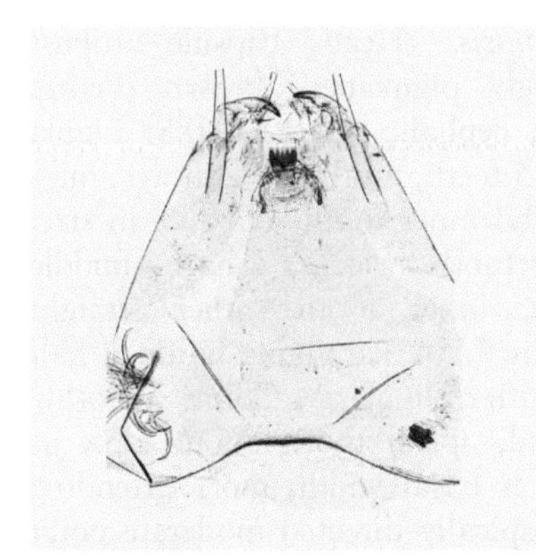

Diagnosis. Head capsule oval, narrow; cephalic index 0.6. Ligula with 5 teeth, tooth row concave to straight, middle section moderately narrowed; basal granulose area forms a narrow stripe (Figure 7.8a,b); paraligula bifid (Figure 7.8b). Pecten hypopharyngis with about 17 subequal teeth. Mandible curved, mola developed as a medium size tooth, inner tooth well developed (Figure 7.8a).

Remarks. Larvae of *Larsia* resemble those of *Natarsia*, in particular with regard to the appearance of ligula and the large molar tooth of the mandibula, however, *Larsia* has well-developed inner mandibular tooth and *Natarsia* has a strongly sclerotized structure on both side of the mentum. Only one larval type of *Larsia* was identified in the Central American material. This resembles the genus *Pentaneura* but can be distinguished from the latter by the broader ligula, concave tooth row, and the shape of basal granulose area.

Ecology and Distribution. This is a worldwide distributed genus, with larvae cold-stenothermic and occurring in several habitats including springs, ditches, streams, small standing waters and the littoral zone of lakes (Cranston and Epler 2013). The group is relatively rare in Central American material, occurring in sediments of lowland lakes.

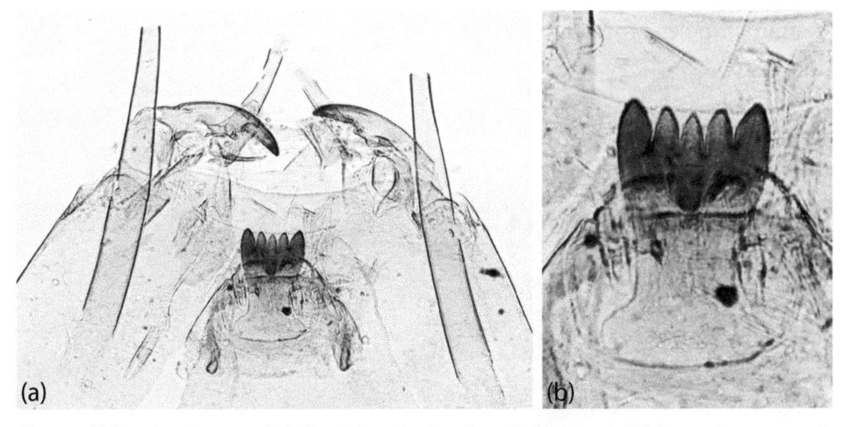

Figure 7.8 *Larsia*. a—detail of head showing ligula, mandible, antenna, and maxillary palp, b—ligula and paraligula.

MACROPELOPIINI TYPE A

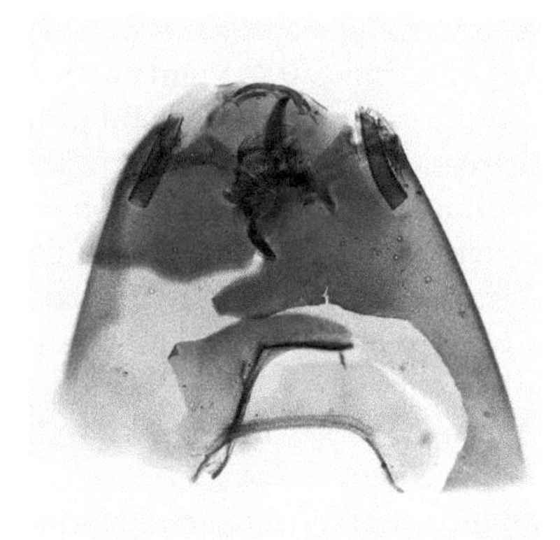

Diagnosis. Head capsule robust strongly pigmented, brown (Figure 7.9a), cephalic index ca 0.8. Ligula with 5 teeth, tooth row concave, middle and inner tooth subequal in size, outer tooth twice as long as middle tooth; inner tooth rather straight (Figure 7.9b). Paraligula bifid, ca half a length of ligula. Mandible smoothly curved, apical tooth about 2.5× as long as basal width, mola extending into apically directed moderate point (Figure 7.9a), small accessory tooth may be present. Mentum with 4 visible dorsomental teeth on each side (there may be an indication of a minute 5th tooth), inner tooth narrow pointy, middle and outer teeth broad and rounded. Antenna slightly longer than mandible.

Remarks. The morphotype most likely belongs to the tribe Macropelopiini. It shows some similarity to *Macropelopia*, but has fewer dorsomental teeth (4–5 pairs). The dorsomentum resembles that of *Apsectrotanypus* (bearing 5 teeth), but there is neither an elongate pale belt across the ventral part of the head, nor visible tentorial lines that are typical for *Apsectrotanypus* (Vallenduuk and Lipinski 2009). There is some similarity to *Brundiniella*, but the dorsomentum lacks the lobe extending to midline, characteristic of *Brundiniella*. A further similar genus, *Alotanypus*, differs from it by the inner ligula teeth pointing strongly outward. All in all, the generic position of this morphotype remains unclear for now.

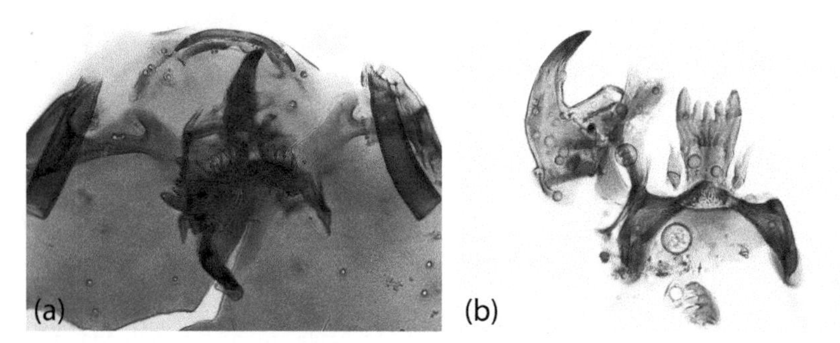

Figure 7.9 Macropelopiini type A. a—detail of head capsule with dark antenna retracted into head, b—ligula, paraligula, mandible, dorsomental plate with 4 teeth.

Ecology and Distribution. Neotropical Macropelopiini are poorly known (Spies et al. 2009), most species have been described from adult specimens only and many do not match any recognized genus (Spies and Reiss 1996). Currently, two Macropelopiini genera are documented in Central America (Spies et al. 2009), but many undescribed species are known that cannot be placed in existing genera. Recently, Silva and Pinho (2020) described a new species from Argentina belonging to genus *Macropelopia*, however, only based on adult males. Macropelopiini type A was rare in Central American lowland lakes.

NILOTANYPUS KIEFFER

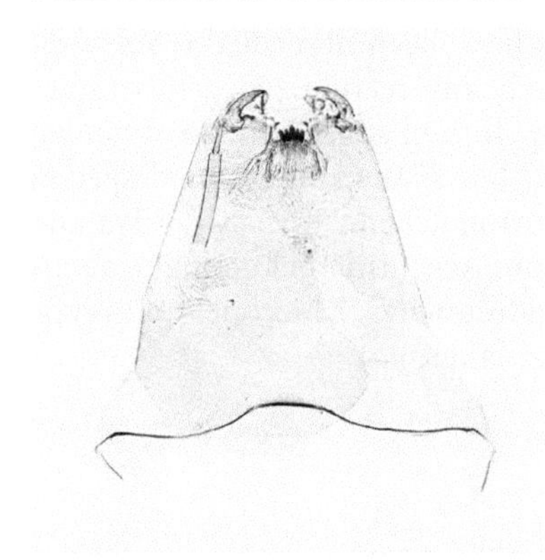

Diagnosis. Head capsule narrow, yellow-brown; cephalic index about 0.45. Ligula with 5 teeth, twice as long as apical width, strongly narrowed in middle (Figure 7.10); basal granulose area of triangular shape. Mandible weakly curved, apical tooth long (about ¼ of length of mandible); inner tooth large, pointed, molar expansion large, pointed (Figure 7.10).

Remarks. Larvae of *Nilotanypus* resemble those of *Labrundinia* in the shape of ligula and granulose basal area, but can be separated by having a much smaller molar expansion, as well as by the smooth head and lack of lateral spinose hump.

Ecology and Distribution. Larvae belonging to the genus *Nilotanypus* inhabit cool, oxygen-rich flowing waters, where they prefer sandy substrates (Cranston and Epler 2013). Owing to their rheobiontic character, the group is rare in Central American lowland lakes and specimens most likely originate from the inlet streams.

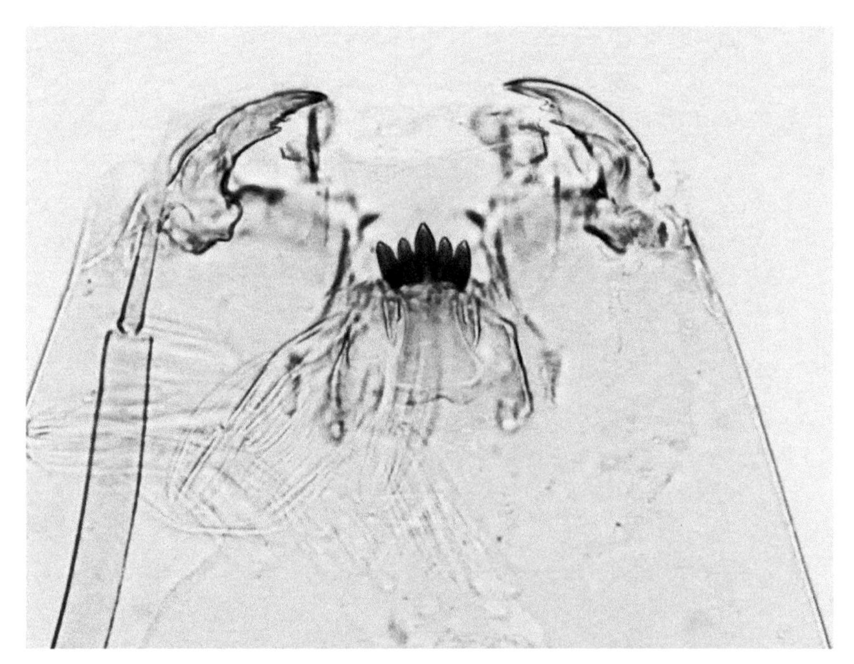

Figure 7.10 *Nilotanypus.* Detail of head with ligula, mandible, and retracted antenna.

PENTANEURA PHILIPPI

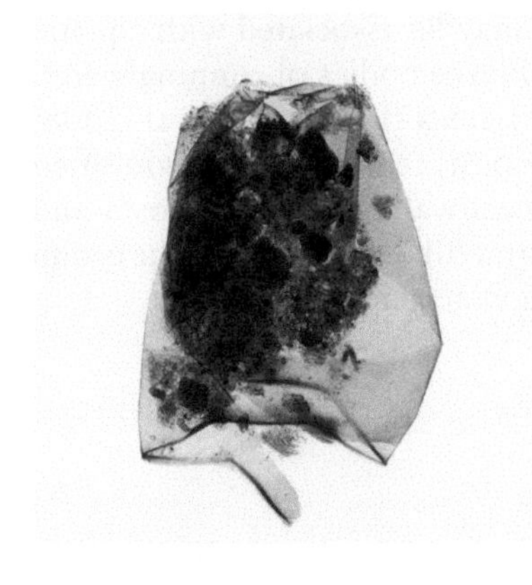

Diagnosis. Head capsule narrow, elongate-oval; cephalic index 0.4–0.7 (Silva and Ferrington 2018, Figure 7.11a). Ligula with 5 teeth of almost equal size, tooth row straight or weakly concave (Figure 7.11a); paraligula bifid, pecten hypopharyngis with 10–13 teeth, with several teeth of variable sizes and shapes. Mandible gradually curved and narrowed toward apex; inner tooth, large, apically directed, rounded, not projecting beyond margin of seta subdentalis (Silva and Ferrington 2018).

Remarks. Living *Pentaneura* larvae can be easily recognized by very long anal tubules and large supraanal setae mounted on dark tubercles. The genus can be distinguished from *Trissopelopia* by the ring organ situated in the apical ⅓ of the basal maxillary palp segment. However, it is hardly distinguishable from other Pentaneurini genera with ligula with straight toothed margin, unless mandible is present.

Ecology and Distribution. *Pentaneura* is a common genus occurring in the Nearctic and Neotropical regions (Ashe and O'Connor 2009), including some islands from the Caribbean region (Ferrington et al.

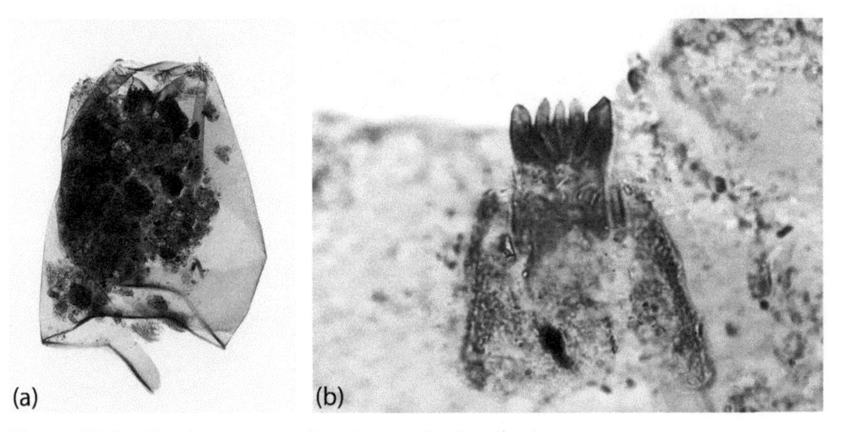

Figure 7.11 *Pentaneura*. a—head capsule, b—ligula.

1993; Silva et al. 2015b). Larval *Pentaneura* inhabit a variety of aquatic systems, from ponds and lakes, where it may be associated with aquatic plants and detritus, to small streams and large bodies of running water, usually living in erosion and depositional areas (Ferrington et al. 2008; Cranston and Epler 2013). In the Neotropical region, *Pentaneura* often has been reported from high elevation headwater streams (Acosta and Prat 2010; Silva and Ferrington 2018; Hamerlík et al. 2018a). The group was rarely recorded in Central American waterbodies.

PROCLADIUS SKUSE

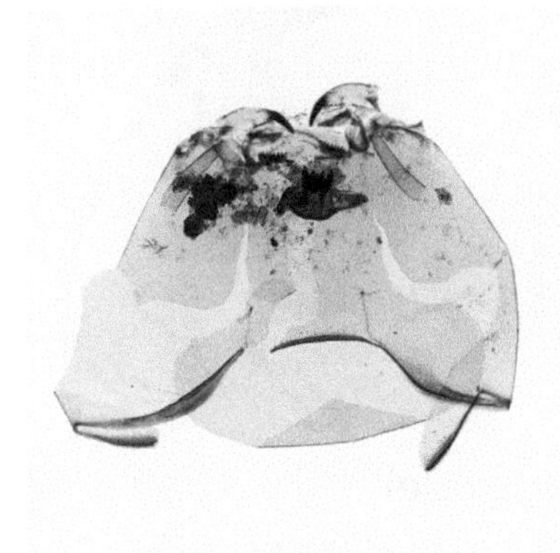

Diagnosis. Large, rounded head capsule, cephalic index 0.80–0.85. Ligula with 5 teeth, median tooth is the shortest, progressively growing toward the outer teeth (Figure 7.12a–c); paraligula multitoothed (Figure 7.12c), pecten hypopharyngis with 10–15 teeth, second row of small teeth can be present. Mandible slender, uniformly curved with a dark apical tooth thrice as long as basal width; mola expanded to large broad, protruding tooth with a rounded apex (Figure 7.12e,g). Mentum with 6–8 pairs of dorsomental teeth (Figure 7.12d,f).

Remarks. Procladius can be distinguished from *Djalmabatista* (ligula with 5 teeth) by dark occipital margin, as opposed to pale in *Djalmabatista*, shape of molar extension on mandible, single and large in *Procladius*, while double in *Djalmabatista*, and the length of antennal blade, which is very short, not exceeding the antennal flagellum in *Procladius*, in contrast to elongate in *Djalmabatista.* Nevertheless, the antennal blade is usually missing in subfossil material and thus it is rather problematic or even impossible to distinguish these two genera.

Ecology and Distribution. Procladius is a widespread, species-rich and diverse genus, most likely all three known subgenera occur in Central America (Spies et al. 2009). Larvae of *Procladius* usually favor muddy

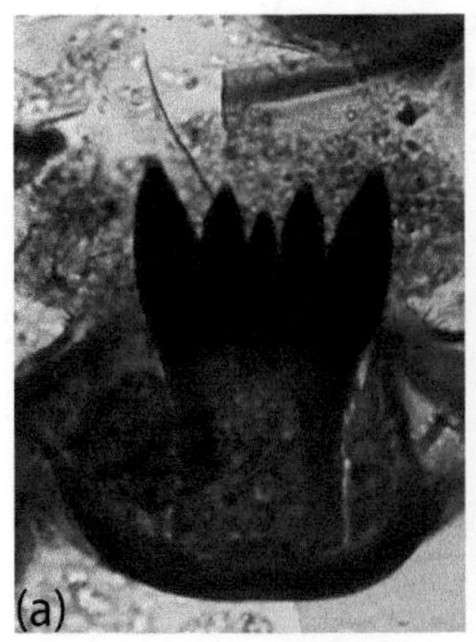

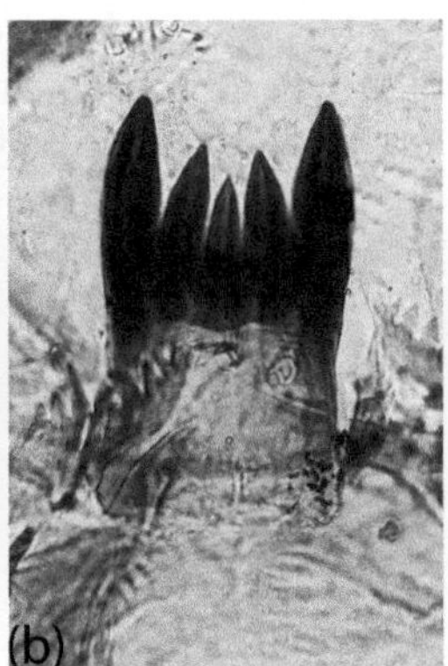

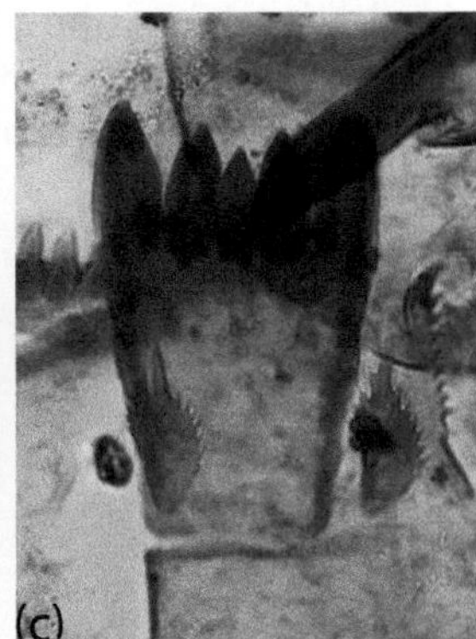

Figure 7.12 *Procladius.* a, b, c—variability in shapes of ligula, d—detail of head capsule with ligula, dorsomental teeth, mandible, and antenna, e—mandible, f—detail of head capsule with ligula, dorsomental teeth, mandible, and antenna (with mandible resembling *Djalmabatista*), g—mandible (similar to *Djalmabatista*).

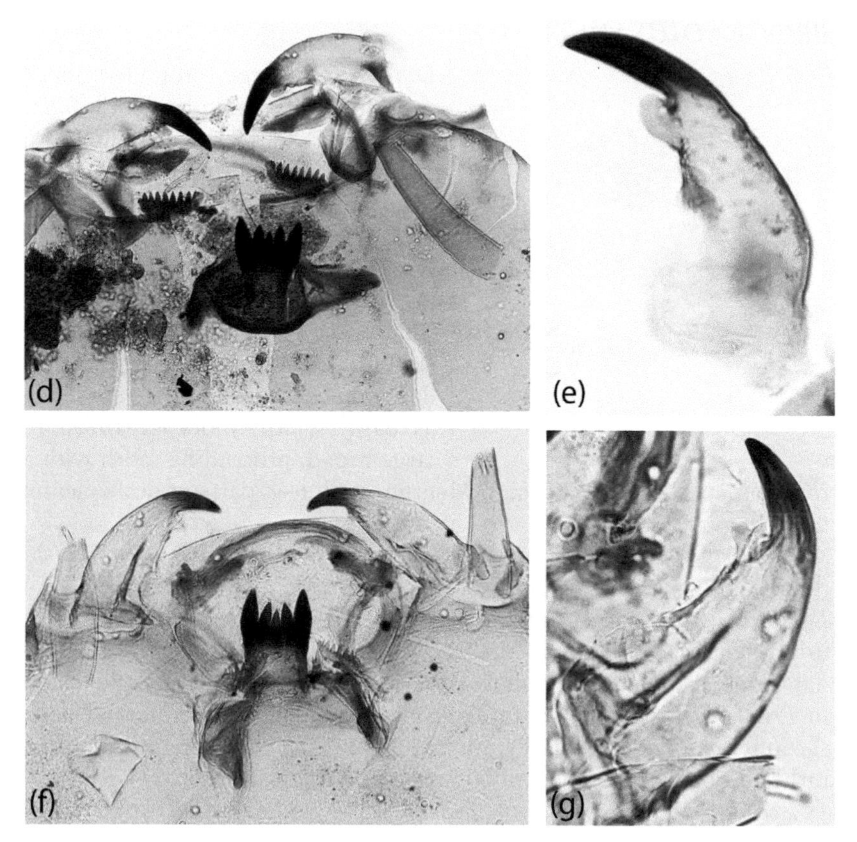

Figure 7.12 *Continued.*

substrata of standing or slow flowing waterbodies, especially ponds and small lakes (Cranston and Epler 2013). Larvae are free living and can tolerate higher organic pollution and consequently low dissolved oxygen content. The group is common in lakes across Central America.

TANYPUS MEIGEN

Diagnosis. Head capsule large, rounded, usually pale with dark occipital margin, cephalic index about 0.95. Ligula pale, slightly convex with five subequal teeth; paraligula large and variable from bifid to multi-branched (Figure 7.13a), pecten hypopharyngis strongly reduced. Mandible with stout base, apical tooth about twice as its basal width, several small, subequal pointy accessory teeth present, mola bilobed apically (Figure 7.13c). Mentum with 5–7 pairs of large dorsomental teeth (Figure 7.13b).

Remarks. The shape of mandible and mentum, the absence of pseudoradula and the reduced pecten hypopharyngis can separate *Tanypus* from all other Tanypodinae. Only one morphotype, *Tanypus stellatus*-type (after Roback 1976), was identified based on the shape of ligula and paraligula.

Ecology and Distribution. This is a widespread and species-rich genus with three named species recorded in Central America (Spies and Reiss, 1996), but presence of more species is likely (Spies et al. 2009). Larval *Tanypus* live in sediments in standing and slow flowing waters, especially in temperate to warm regions, since they can tolerate high nutrient and salinity levels (Cranston and Epler 2013). Remains of *Tanypus stellatus*-type were common in lowland lakes of El Salvador and Guatemala.

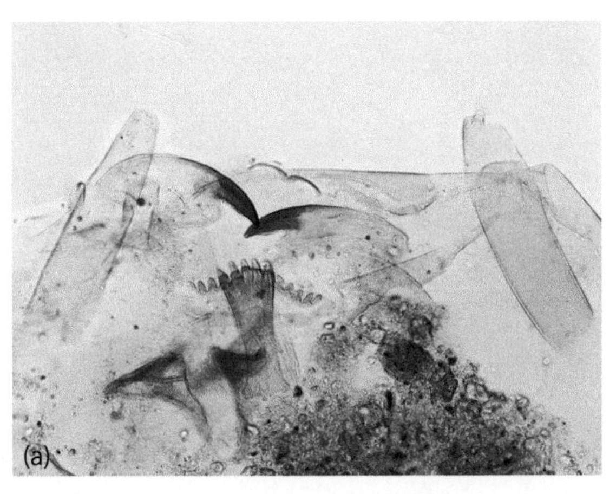

Figure 7.13 a—detail of head capsule showing ligula, dorsomentum, mandible and antenna.

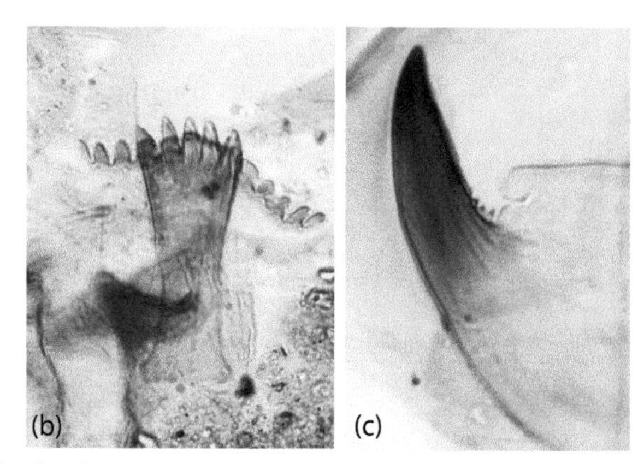

Figure 7.13 b—ligula and dorsomentum, c—detail of mandible with accessory teeth.

References

Acosta R, Prat N (2010) Chironomid assemblages in high altitude streams of the Andean region of Peru. *Fundamental and Applied Limnology* 177(1): 57–79.

Albuquerque FS, Benito B, Beier P, Assunção-Albuquerque MJ, Cayuela L (2015) Supporting underrepresented forests in Mesoamerica. *Natureza e Conservação* 13(2): 152–158.

Alcocer J, Lugo A, Estrada S, Ubeda M, Escobar E (1993) La macrofauna bentónica de los axalapazcos mexicanos. *Actas del Vi Congreso Español de Limnología* 33: 409–415.

Andersen T, Contreras Ramos A, Spies M (2000) Chironomidae (Diptera), pp. 581–591. In: Llorente B, Gonzales SJE, Papavero N (eds). *Biodiversidad, Taxonomía y Biogeografía de Artrópodos de México: Hacia una Síntesis de Su Conocimiento.* Vol. II. Universidad Nacional Autónoma de México, Mexico City, 676 pp.

Andersen T, Kristoffersen L (1998) New species of *Xestochironomus* Sublette and Wirth (Chironomidae: Chironominae) from Chile and Costa Rica. *Journal of the Kansas Entomological Society* 71: 296–303.

Andersen T, Mendes HF (2004) *Irisobrillia longicosta* Oliver, 1985 (Diptera: Chironomidae: Orthocladiinae) taken in south Brazil. *Biota Neotropica* 4(2): 1–5.

Andersen T, Mendes HF, Pinho LC (2015) *Mariambera*, a new genus of Orthocladiinae from Brazil (Insecta: Diptera, Chironomidae). *Studies on Neotropical Fauna and Environment* 50(1): 24–30.

Andersen T, Sæther OA, Cranston PS, Epler JH (2013) The larvae of Orthocladiinae (Diptera: Chironomidae) of the Holarctic region – Keys and diagnoses. In: Andersen T, Cranston PS, Epler JH (eds). The Larvae of Chironomidae (Diptera) of the Holarctic Region – Keys and Diagnoses. *Insects Systems Evolution, Suppl.* 66: 189–386.

Armitage PD, Cranston PS, Pinder LCV (1995) *The Chironomidae: Biology and Ecology of Non-Biting Midges.* Chapman & Hall, London, Glasgow, Weinheim, New York, Tokyo, Melbourne & Madras, xii + 572 pp.

Ashe P, Cranston PS (1990) Family Chironomidae, pp. 113–335. In: Soós Á, Papp L (eds). *Catalogue of Palaearctic Diptera.* Vol. 2. Psychodidae – Chironomidae. Akadémiai Kiadó, Budapest.

Ashe P, O'Connor JP (2009) *A World Catalogue of Chironomidae (Diptera). Part 1. Buchonomyiinae, Chilenomyiinae, Podonominae, Aphroteniinae, Tanypodinae, Usambaromyiinae, Diamesinae, Prodiamesinae and Telmatogetoninae.* Irish Biogeographical Society & National Museum of Ireland, Dublin, Ireland.

Bitušík P (2000) *Handbook for Identification of Chironomid Larvae (Diptera: Chironomidae) of Slovakia. Part I. Buchonomyinae, Diamesinae, Prodiamesinae and Orthocladiinae.* Vydavateľstvo Technickej Univerzity, Zvolen, Slovakia (in Slovak).

Bitušík P, Hamerlík L (2014) *Identification key for Chironomidae Larvae of Slovakia. Part 2. Tanypodinae.* Belianum, Vydavateľstvo Univerzity Mateja Bela v Banskej Bystrici, Slovakia, 96 pp. [in Slovak].

Borkent A (1984) The systematics and phylogeny of the *Stenochironomus* complex (*Xestochironomus*, *Harrisius*, and *Stenochironomus*). (Diptera: Chironomidae). *The Memoirs of the Entomological Society of Canada* 128: 1–269.

Brooks SJ, Langdon PG, Heiri O (2007) The identification and use of Palaearctic Chironomidae larvae in palaeoecology. *QRA Technical Guide No. 10.* QRA, London, 276 pp.

Brundin L (1983) The larvae of Podonominae (Diptera: Chironomidae) of the Holarctic region – Keys and diagnoses. *Entomologica Scandinavica, Suppl.* 19: 23–31.

Chacon CM (2005) Fostering conservation of key priority sites and rural development in Central America: The role of private protected areas. *Parks* 15: 39–47.

Cione AL, Gasparini GM, Soibelzon E, Soibelzon LH, Tonni EP (2015) *The Great American Biotic Interchange: A South American Perspective.* Springer Netherlands.

Cranston PS (1982) *A Key to the Larvae of the British Orthocladiinae (Chironomidae).* Freshwater Biological Association, Scientific Publication 45. The Ferry Hause, Ambleside, Cumbria LA22 oLP. 152 pp.

Cranston PS (1995) Introduction to the Chironomidae, pp. 1–7. In: Armitage PD, Cranston PS, Pinder LCV (eds). *The Chironomidae: The Biology and Ecology of Non-Biting Midges.* Chapman & Hall, New York.

Cranston PS (1995) Morphology, pp. 11–30. In: Armitage PD, Cranston PS, Pinder LCV (eds). *The Chironomidae: The Biology and Ecology of Non-Biting Midges.* Chapman & Hall, London, Glasgow, Weinheim, New York, Tokyo, Melbourne & Madras, xii + 572 pp.

Cranston PS (2010) Electronic guide to the Chironomidae of Australia. http://chirokey.skullisland.info/. [30 December 2019].

Cranston PS (2013) The larvae of the Holarctic Chironomidae (Diptera) – Morphological terminology and keys to subfamilies. In: Andersen T, Cranston PS, Epler JH (eds). The Larvae of Chironomidae (Diptera) of the Holarctic Region – Keys and Diagnoses. *Insect Systematics & Evolution, Suppl.* 66: 13–23.

Cranston PS, Epler J (2013) The larvae of Tanypodinae (Diptera: Chironomidae) of the Holarctic region – Keys and diagnoses. In: Andersen T, Cranston PS, Epler JH (eds). The Larvae of

Chironomidae (Diptera) of the Holarctic Region – Keys and Diagnoses. *Insect Systematics and Evolution, Suppl.* 66: 39–136. Lund, Sweden

Cranston PS, Hardy NB, Morse GE (2012) A dated molecular phylogeny for the Chironomidae (Diptera). *Systematic Entomology* 37: 172–188.

Cranston PS, Hardy NB, Morse GE, Puslednik L, McCluen SR (2010) When molecules and morphology concur: The 'Gondwanan' midges (Diptera: Chironomidae). *Systematic Entomology* 35(4): 636–648.

Cranston PS, Nolte U (1996) *Fissimentum*, a new genus of drought-tolerant Chironomini (Diptera: Chironomidae) from the Americas and Australia. *ENT News* 107: 1–15.

Cranston PS, Oliver DR, Sæther OA (1983) The larvae of Orthocladiinae (Diptera, Chironomidae) of the Holarctic region – Keys and diagnoses. *Entomologica Scandinavica, Suppl.* 19: 149–291.

Cranston PS, Oliver DR, Sæther OA (1989) Keys and diagnoses of the adult males of the subfamily Orthocladiinae (Diptera, Chironomidae). *Entomologica Scandinavica, Suppl.* 34: 165–352.

Dantas GPS, Pinheiro MPG, Hamada N (2020) An unusual new species of *Pentaneura* Philippi (Diptera: Chironomidae) from northeastern Brazil, with an emended diagnosis to the genus. *Zootaxa* 4786 (1): 81–92

Donato M, Paggi AC (2005) A new Neotropical species of the genus *Metriocnemus* van der Wulp (Chironomidae: Orthocladiinae) from *Eryngium* L. (Apiaceae) phytotelmata. *Zootaxa* 1050(1): 1–14.

Epler JH (2001) Identification manual for the larval Chironomidae (Diptera) of North and South Carolina: A guide to the taxonomy of the midges of the Southeastern United States, including Florida. Special Publication SJ2001-SP13 North Carolina, Department of Environment and Natural Resources, Raleigh, NC, and St. Johns River Water Management District, Palatka, FL, 526 pp.

Epler JH (2014) *Identification Guide to the Larvae of the Tribe Tanytarsini (Diptera: Chironomidae) in Florida.* Florida Department of Environmental Protection, Tallahassee, 77 pp.

Epler JH (2017) An annotated preliminary list of the Chironomidae (Diptera) of Zurquí, Costa Rica. *Chironomus Journal of Chironomidae Research* 30: 4–18.

Epler JH, Ekrem T, Cranston PS (2013) The larvae of Chironominae (Diptera: Chironomidae) of the Holarctic region – Keys and diagnoses. In: Andersen T, Cranston PS, Epler JH (eds). The Larvae of Chironomidae (Diptera) of the Holarctic Region – Keys and Diagnoses. *Insects Systems Evolution Suppl.* 66: 387–556.

Ferrington LC Jr (2008) Global diversity of non-biting midges (Chironomidae, Insecta, Diptera) in freshwater. *Hydrobiologia* 595(1): 447–455.

Ferrington LC Jr, Berg MB, Coffman WP (2008) Chironomidae, pp. 847–989. In: Merritt RW, Cummins KW, Berg MB (eds). *An Introduction to the Aquatic Insects of North America*. Kendall/Hunt Publishing Co., Dubuque, LA.

Ferrington LC Jr, Buzby KM, Masteller EC (1993) Composition and temporal abundance of Chironomidae emergence from a tropical rainforest stream at el Verde, Puerto Rico. *Journal of Kansas Entomological Society* 66: 167–180.

Fittkau EJ, Roback SS (1983) The larvae of Tanypodinae (Diptera: Chironomidae) of the Holarctic region – Keys and diagnoses. *Entomologica Scandinavica, Suppl.* 19: 33–110.

Frey DG (1976) Interpretation of Quaternary paleolimnology from Cladocera and midges and prognosis regarding usability of other organisms. *Canadian Journal of Zoology* 54(12): 2208–2226.

Fusari LM, Roque FDO, Hamada N (2014) Systematics of *Oukuriella* Epler, 1986, including a revision of the species associated with freshwater sponges. *Insect Systematics and Evolution* 45(2): 117–157.

Hamerlík L, Silva FL (2018) First record of the genus *Heterotrissocladius* (Chironomidae: Orthocladiinae) from the Neotropical region. *Chironomus Journal of Chironomidae Research* 31(31): 43–46.

Hamerlík L, Silva FL, Jacobsen D (2018a) Chironomidae (Insecta: Diptera) of Ecuadorian high-altitude streams: A survey and illustrated key. *Florida Entomologist* 101(4): 663–676.

Hamerlik L, Silva FL, Massaferro J. An illustrated guide of subfossil Chironomidae (Insecta: Diptera) from waterbodies of Central America and the Yucatan Peninsula. *Journal of Paleolimnology* (submitted, not yet accepted).

Hamerlík L, Silva FL, Wojewodka M (2018b) Sub-fossil Chironomidae (Diptera) from lake sediments in Central America: A preliminary inventory. *Zootaxa* 4497(4): 559–572.

Hofmann W (1971) Zur Taxonomie und Palökologie subfossiler Chironomiden (Dipt.) in Seesedimenten. *Archiv fur Hydrobiologie Beih* 6: 1–50.

Janssens De Bisthoven L, Timmermans KR, Ollevier F (1992) The concentration of cadmium, lead, copper and zinc in *Chironomus* gr. *thummi* larvae (Diptera: Chironomidae) with deformed versus normal menta. *Hydrobiologia* 239(3): 141–149.

Kowalyk HE (1985) The larval cephalic setae in the Tanypodinae (Diptera: Chironomidae) and their importance in generic determinations. *Canadian Entomologist* 117(1): 67–106.

Krell FT (2004) Parataxonomy vs. taxonomy in biodiversity studies – Pitfalls and applicability of 'morphospecies' sorting. *Biodiversity and Conservation* 13(4): 795–812.

Lin XL, Stur E, Ekrem T (2018) Molecular phylogeny and temporal diversification of Tanytarsus van der Wulp (Diptera: Chironomidae) support generic synonymies, a new classification and centre of origin. *Systematic Entomology* 43(4): 659–677.

Moller Pillot HKM (2009) *Chironomidae Larvae of the Netherlands and Adjacent Lowlands: Biology and Ecology of the Chironomini.* KNNV Publishing, Zeist, the Netherlands, 174.

Moller Pillot HKM (2013) *Chironomidae Larvae of the Netherlands and Adjacent Lowlands: Biology and Ecology of the Aquatic Orhocladiinae.* KNNV Publishing, Zeist, the Netherlands 314 pp.

Morrone JJ (2014) Cladistic biogeography of the Neotropical region: Identifying the main events in the diversification of the terrestrial biota. *Cladistics* 30(2): 202–214.

Nazarova LB (2000) A point of view on chironomid deformities investigation. *Chironomus Journal of Chironomidae Research* 13(14): 7–8.

Oliver DR (1971) Life history of the Chironomidae. *Annual Review of Entomology* 12(1): 211–230.

Parise AG, Pinho LC (2016) A new species of *Stenochironomus* Kieffer, 1919 from the Atlantic Rainforest in southern Brazil (Diptera: Chironomidae). *Aquatic Insects* 37(1): 1–7.

Pérez L, Bugja R, Massaferro J, Steeb P, Geldern R, Frenzel P, Brenner M, Scharf B, Schwalb A (2010) Post-Columbian environmental history of Lago Petén Itzá. *Guatemala Revista Mexicana de Ciencias Geológicas* 27: 490–507.

Pérez L, Lorenschat J, Massaferro J, Pailles C, Sylvestre F, Hollwedel W, Brandorff G, Brenner M, Gerald I, Lozano S, Scharf B, Schwalb A (2013) Bioindicators of climate and trophic state in lowland and highland aquatic ecosystems of the Northern Neotropics. *Revista de Biología Tropical* 61(2): 603–644.

Pinder LCV, Reiss F (1983) The larvae of Chironominae (Diptera: Chironomidae) of the Holarctic region – Keys and diagnoses. *Entomologica Scandinavica, Suppl.* 19: 293–435.

Rieradevall M, Brooks SJ (2001) An identification guide to subfossil Tanypodinae larvae (Insecta: Diptera: Chrironomidae) based on cephalic setation. *Journal of Paleolimnology* 25: 8–99.

Roback SS (1976) The immature chironomids of the Eastern United States II. Tanypodinae: Tanypodini. *Proceedings of the Academy of Natural Sciences of Philadelphia* 128: 55–88.

Roque FO, Siqueira T, Trivinho-Strixino S (2005) Occurrence of chironomid larvae living inside fallen-fruits in Atlantic Forest streams, Brasil. *Entomología y Vectores* 12(2): 275–282.

Roque FO, Trivinho-Strixino S (2008) Four new species of *Endotribelos* Grodhaus, a common fallen fruit-dwelling chironomid genus in Brazilian streams (Diptera: Chironomidae: Chironominae). *Studies on Neotropical Fauna and Environment* 43(3): 191–207.

Sæther OA (1975) Nearctic and Palearctic *Heterotrissocladius* (Diptera, Chironomidae). *Bulletin of the Fisheries Reseanth Board of Canada* 193: 1–67.

Sæther OA (1980) Glossary of Chironomid morphology terminology (Diptera: Chironomidae). *Entomologica Scandinavica: Supplement* 14: 1–51.

Sæther OA (1989) *Metriocnemus* van der Wulp: A new species and a revision of species described by Meigen, Zetterstedt, Staeger, Holmgren Lundstrom and Strenzke (Diptera: Chironomidae). *Entomologica Scandinavica* 19: 393–430.

Sæther OA (2000) Phylogeny of the subfamilies of Chironomidae (Diptera). *Systematic Entomology* 25(3): 393–403.

Sæther OA, Wang X (1995) Revision of the genus *Paraphaenocladius* Thienemann, 1924 of the world (Diptera: Chironomidae, Orthocladiinae). *Entomologica Scandinavica: Supplement* 48: 1–69.

Sanchez-Azofeifa A, Powers JS, Fernandes GW, Quesada M (eds.) (2014) *Tropical Dry Forests in the Americas: Ecology, Conservation, and Management*. CRC Press, Boca Raton, FL.

Schmid PE (1993) *A Key to the Larval Chironomidae and Their Instars from the Austrian Danube Region Streams and Rivers. Part 1. Diamesinae, Prodiamesinae and Orthocladiinae*. Federal Institute for Water Quality, Wien.

Silva FL, Ekrem T (2016) Phylogenetic relationships of nonbiting midges in the subfamily Tanypodinae (Diptera: Chironomidae) inferred from morphology. *Systematic Entomology* 41(1): 73–92.

Silva FL, Ekrem T, Fonseca-Gessner AA (2015a) Out of South America: Phylogeny of non-biting midges in the genus *Labrundinia* (Diptera: Chironomidae) suggests multiple dispersal events to Central and North America. *Zoologica Scripta* 44(1): 59–71.

Silva FL, Farrell BD (2017) Non-biting midges (Diptera: Chironomidae) research in South America: Subsidizing biogeographic hypotheses. *Annales de Limnologie—International Journal of Limnology* 53: 111–128.

Silva FL, Ferrington Jr LC (2018) Systematics of the New World genus *Pentaneura* Phillipi (Diptera: Chironomidae: Tanypodinae): Historical review, new species and phylogeny. *Zoologischer Anzeiger* 271: 1–31.

Silva FL, Fonseca-Gessner AA (2009) The immature stages of *Labrundinia tenata* (Diptera: Chironomidae: Tanypodinae) and redescription of the male. *Zoologia* 26: 541–546.

Silva FL, Fonseca-Gessner AA, Ekrem T (2011) Revision of *Labrundinia maculata* Roback, 1971, a new junior synonym of *L. longipalpis* (Goetghebuer, 1921) (Diptera: Chironomidae: Tanypodinae). *Aquatic Insects* 33(4): 293–303.

Silva FL, Fonseca-Gessner AA, Ekrem T (2014) A taxonomic revision of genus *Labrundinia* Fittkau, 1962 (Diptera: Chironomidae: Tanypodinae). *Zootaxa* 3769: 1–185.

Silva FL, Oliveira CSN (2016) *Tanypus urszulae*, a new Tanypodinae (Diptera: Chironomidae) from the Neotropical region. *Zootaxa* 4178(4): 593–600.

Silva FL, Pinho LC (2020) *Macropelopia* (*Macropelopia*) *patagonica*, a new Tanypodinae (Diptera: Chironomidae) from the Patagonian Andes. *Zootaxa* 4731(4): 574–580.

Silva FL, Pinho LC, Wiedenbrug S, Dantas GP, Siri A, Andersen T, Trivinho-Strixino S (2018) Family Chironomidae, pp. 661–700. In: *Thorp and Covich's Freshwater Invertebrates.* Academic Press, Cambridge, MA.

Silva FL, Wiedenbrug S (2014) Integrating DNA barcodes and morphology for species delimitation in the *Corynoneura* group (Diptera: Chironomidae). *Bulletin of Entomological Research* 104(1): 65–78.

Silva FL, Wiedenbrug S, Farrell B (2015b) A preliminary survey of the non-biting midges (Diptera: Chironomidae) of the Dominican Republic. *Chironomus Journal of Chironomidae Research* 28(28): 12–19.

Siri A, Donato M (2015) Phylogenetic analysis of the tribe Macropelopiini (Chironomidae: Tanypodinae): Adjusting homoplasies. *Zoological Journal of the Linnean Society* 174(1): 74–92.

Spies M, Andersen T, Epler JH, Watson CN (2009) Chironomidae (non-biting midges), pp. 437–480. In: Brown BV, Borkent A, Cumming JM, Wood DM, Woodley NE, Zumbado M (eds). *Manual of Central American Diptera.* NRC Research Press, Ottawa.

Spies M, Reiss F (1996) Catalog and bibliography of Neotropical and Mexican Chironomidae. *Spixiana: Supplement* 22: 61–119.

Spies M, Sæther OA (2013) Fauna europaea: Chironomidae. In: Pape T, Beuk P (eds). *Fauna Europaea: Diptera, Nematocera: Fauna Europaea Version 2018.08.* http://www.faunaeur.or.

Sublette JE, Sasa M (1994) Chironomidae collected in onchocerciasis endemic areas in Guatemala (Insecta, Diptera). *Spixiana: Supplement* 20: 1–60.

Trivinho-Strixino S (2014) Ordem Diptera. Família Chironomidae. Guia de identificação de larvas, pp. 457–660. In: Hamada N, Nessimian JL, Querino RB (eds). *Insetos Aquático na Amazônia Brasileira: Taxonomia, Biologia e Ecologia*, Editora do Instituto Nacional de Pesquisas Da Amazônia, Manaus.

Trivinho-Strixino S, Silva FL, Oliveira CSN (2013) *Tapajos cristinae* n. gen., n. sp. (Diptera: Chironomidae: Chironominae) from Neotropical region. *Zootaxa* 3710: 395–399.

Trivinho-Strixino S, Strixino G (2008) A new species of *Pelomus* Reiss, 1989 (Diptera: Chironomidae from Southeastern Brazil, with the Description of Immature Stages. Boletim Do Museu Municipal De Funchal, Suplemento) from southeastern Brazil, with the description of immature stages. *Boletim do Museu Municipal de Funchal, Suplemento* 13: 223–231.

Trivinho-Strixino S, Wiedenbrug S, Silva FL (2015) New species of *Tanytarsus* van der Wulp (Diptera: Chironomidae: Tanytarsini) from Brazil. *European Journal of Environmental Sciences* 5(1): 92–100.

Turcotte P, Harper PP (1982) The macroinvertebrate fauna of a small Andean stream. *Freshwater Biology* 12: 411–419.

Vallenduuk H, Lipinski A (2009) Neglected and new characters in Chironomidae: Tanypodinae (larvae). *Lauterbornia* 68: 83–93.

Vinogradova EM, Riss HW (2007) Chironomids of the Yucatán Peninsula. *Chironomus Journal of Chironomidae Research* 20(20): 32–35.

Warwick WF (1980) Chironomid (Diptera) responses to 2800 years of cultural influence. A paleolimnogical study with special reference to sedimentation, eutrophication, and contamination processes. *Canadian Journal of Entomology* 112: 1193–1238.

Warwick WF (1985) Morphological abnormalities in Chironomidae (Diptera) larvae as measures of toxic stress in freshwater ecosystems: Indexing antennal deformities in Chironomus Meigen. *Canadian Journal of Fisheries and Aquatic Sciences* 42(12): 1881–1914.

Watson CN, Heyn MW (1992) A preliminary survey of the Chironomidae (Diptera) of Costa Rica, with emphasis on the lotic fauna. *Netherland Journal of Aquatic Ecology* 26(2–4): 257–262.

Werle SF, Smith DG, Klekowski E (2004) Life in crumbling clay: The biology of *Axarus* species (Diptera: Chironomidae) in the Connecticut River. *Northeastern Naturalist* 11(4): 443–459.

Whitmore TJ, Riedinger-Whitmore MA (2014) Topical advances and recent studies in paleolimnological research. *Journal of Limnology* 73(s1): 149–160.

Wu J, Porinchu DF, Campbell NL, Mordecai TM, Alden EC (2019a) Holocene hydroclimate and environmental change inferred from a high-resolution multi-proxy record from Lago Ditkebi, Chirripó National Park, Costa Rica. *Palaeogeography, Palaeoclimatology, Palaeoecology* 518: 172–186.
Wu J, Porinchu DF, Horn SP (2017) A Chironomid-based reconstruction of late-Holocene climate and environmental change for southern Pacific Costa Rica. *Holocene* 27(1): 73–84.
Wu J, Porinchu DF, Horn SP (2019b) Late Holocene hydroclimate variability in Costa Rica: Signature of the terminal classic drought and the Medieval Climate Anomaly in the northern tropical Americas. *Quaternary Science Reviews* 215: 144–159.
Wu J, Porinchu DF, Horn SP, Haberyan KA (2015) The modern distribution of chironomid sub-fossils (Insecta: Diptera) in Costa Rica and the development of a regional chironomid-based temperature inference model. *Hydrobiologia* 742(1): 107–127.

List of Morphotypes Illustrated in the Book

Ablabesmyia janta-type
Ablabesmyia monilis-type
Apedilum sp.
Axarus sp.
Beardius type Chanmico
Beardius type Ipala
Brillia sp.
Bryophaenocladius/Gymnometriocnemus sp.
Chironomus anthracinus-type
Chironomus plumosus-type
Chironomus type Salto Grande
Cladopelma lateralis-type
Cladotanytarsus mancus-type
Cladotanytarsus type A
Clinotanypus sp.
Coelotanypus concinnus-type
Corynoneura lobata-type
Cricotopus type A
Cricotopus type Atitlan
Cricotopus type B
Cricotopus type C
Cricotopus type I
Cricotopus type Magdalena
Cryptochironomus sp.
Dicrotendipes nervosus-type A
Dicrotendipes nervosus-type B
Dicrotendipes nervosus-type C
Dicrotendipes notatus-type
Djalmabatista sp.
Endochironomus albipennis-type
Endotribelos albatum-type
Endotribelos grodhausi-type
Endotribelos hesperium-type
Eukiefferiella claripennis-type
Eukiefferiella devonica-type
Eukiefferiella gracei-type
Fissimentum desiccatum-type
Fittkauimyia sp.

Goeldichironomus amazonicus-type
Goeldichironomus carus-type
Goeldichironomus holoprasinus-type
Goeldichironomus type Olomega
Harnischia complex type A
Heterotrissocladius marcidus-type
Labrundinia jasoni-type
Labrundinia paulae-type
Labrundinia type G
Labrundinia type Y
Labrundinia virescens-type
Larsia sp.
Lauterborniella sp.
Limnophyes/Paralimnophyes sp.
Macropelopiini type A
Metriocnemus eurynotus-type
Micropsectra type Magdalena
Microtendipes pedellus-type
Microtendipes rydalensis-type
Nanocladius rectinervis-type
Nilotanypus sp.
Nilothauma sp.
Orthocladiinae type A
Orthocladius (*Euorthocladius*) sp.
Oukuriella pinhoi-type
Parachironomus varus-type A
Parachironomus varus-type B
Parakiefferiella bathophila-type
Paralauterborniella sp.
Parametriocnemus/Paraphaenocladius sp.
Paratanytarsus penicillatus-type
Paratendipes nudisquama-type
Parochlus sp.
Pelomus psammophilus-type
Pentaneura sp.
Polypedilum beckae-type
Polypedilum fallax-type
Polypedilum nubeculosum-type
Polypedilum tripodura-type
Polypedilum type A
Procladius sp.
Psectrocladius psilopterus-type
Psectrocladius sordidellus-type
Pseudochironomini type A
Pseudochironomus prasinatus-type

Index

Bold page numbers indicate figures.

For Product Safety Concerns and Information please contact
our EU representative GPSR@taylorandfrancis.com Taylor & Francis
Verlag GmbH, Kaufingerstraße 24, 80331 München, Germany

For Product Safety Concerns and Information please contact our EU representative GPSR@taylorandfrancis.com Taylor & Francis Verlag GmbH, Kaufingerstraße 24, 80331 München, Germany

T - #0238 - 160425 - C198 - 234/156/9 - PB - 9780367076061 - Gloss Lamination